AI and the Art of Teaching

AI will change How we teach, not Why we teach

LaShae Lee, PhD

Illustrations by Christine Lee

Human Learning by Design™

Human Learning by Design™

A series exploring learning, leadership, and human development in an age shaped by artificial intelligence.

AI and the Art of Teaching
AI and the Art of Teaching Workbook

Forthcoming titles in the Human Learning by Design™ series include:
Designing Confidence
Designing Inquiry
Designing Judgment
Designing Learning Environments
Designing Professional Identity
Designing Schools
Designing Support

The developmental Human Learning by Design™ series include:
Becoming Human: Early Childhood
Becoming Human: Middle Childhood
Becoming Human: Adolescence
Becoming Human: Early Adulthood
Becoming Human: Adulthood
Becoming Human: Later Adulthood

Related titles in the Human Learning by Design™ collection:
Raising Humans by Design
Courage Under Algorithm for Teens
Judgment Over Speed: Leadership in an AI Workplace

ISBN: 978-1-972397-00-8 (paperback)
ISBN: 978-1-972397-01-5 (e-book)

Illustrations and cover design by Christine Lee

First Edition, 2026

This book is intended for educational and informational purposes only. The views expressed are those of the author and do not necessarily reflect the policies or positions of any affiliated institutions.

Published and printed in the United States of America

CONTENTS

DEDICATION

This book is dedicated to the people who taught me long before I ever started teaching and helped me grow thereafter.

To my mum, Trudy. You were my first teacher of steadiness. Before I knew theory, I knew your presence. You taught me that learning begins with safety, that correction can be gentle, and that high expectations and compassion belong together. You modeled quiet resilience and daily dignity. "Yesterday is history, tomorrow is a mystery, today is a gift." In those words, you taught me attention and gratitude — to shape what can be shaped now. When I hesitated, you named possibility. When I doubted, you spoke confidence. I first learned belonging with you. I miss you with every breath.

To my husband, Rob. You have walked beside me since high school, witnessing the unseen drafts of who I was becoming — the doubt, the fatigue, the unfinished thinking. You steadied me and protected space for reflection. You tested my arguments and pressed for clarity. When I questioned whether I was enough, you reflected back who I was becoming. Your confidence became my courage.

To my best friend, Carrie. You have been my mirror and compass. You laughed when the weight felt heavy and asked harder questions when I reached for easier answers. You celebrated ideas before they were polished and reminded me that depth matters more than appearance. You refused to let me shrink. Through honest conversation and steady encouragement, you helped me trust my voice.

You three were my most profound teachers – not because you stood behind lecterns, but because you embodied what you hoped I would become. You taught me that presence shapes identity, that trust cultivates growth, and that consistent love forms the architecture of learning.

This book carries your fingerprints in every chapter.

AI and the Art of Teaching

PROLOGUE

The Questions No Algorithm Could Answer

I have been asked many questions as an educator that stopped me short. Some challenged my professional expertise. Others tugged at my moral core.

In one high school classroom of incarcerated students, a young man asked me, "What would you do if you walked into a room and found your little sister being raped?" The question was not meant to provoke. It was asked quietly, sincerely – by a student trying to understand my humanity as much as my authority.

In a middle school classroom, another student, clearly hurt, asked why I thought he would give his friends AIDS by sharing needles "when he was such a nice guy." His understanding of disease transmission was rooted in intention rather than biology. That question, too, was genuine. And painful.

I answered both questions with the best professionalism and humanity I could muster in the moment. I have answered them far more eloquently in my head since. What I understand now is that these questions were not about content. They were about meaning, identity, and trust. They revealed how differently my students and I had experienced the world – and how much teaching depends on navigating those differences with care.

Then, years later, a different kind of question stopped me in a new way.

It came at the end of a lesson, during that familiar moment when attention thins and bodies begin to shift. A student raised their hand – not urgently, not defiantly.

"Can I ask something that's not on the worksheet?" I nodded.

"If AI can answer anything," they asked, "why do we still need school?"

The room went quiet. Not because the question was disruptive, but because it landed somewhere deeper than curriculum. It wasn't a challenge. It was curiosity mixed with uncertainty – the kind that doesn't want reassurance, only honesty.

In that moment, I did not come up with a follow-up question to ask and every prepared response felt insufficient. I could have talked about jobs, or critical thinking, or social skills. All of that would have been true – and still incomplete. Because the question underneath was not really about school. It was this: If machines can produce answers, what is learning for? And where does that leave us?

That question changed how I understood teaching.

Education has always adapted to change. Chalkboards replaced slates. Calculators replaced slide rules. The internet replaced encyclopedias. Each shift brought panic, predictions of decline, and calls to return to 'real learning.' And yet, teaching endured.

Artificial intelligence feels different – not because it stores information, but because it produces language, solutions, and explanations that sound like thinking. AI unsettles a belief we have quietly held for generations: that knowing things is what makes us valuable.

But teaching is not about knowing the most. It is about helping others make sense of the world and their place within it.

This book is not a defense against artificial intelligence. It is not a rejection of technology. It is an examination of what AI reveals – about learning, about schooling, and about the quiet misunderstandings we have tolerated for far too long. This book examines how teaching must be intentionally designed within conditions shaped by intelligent machines.

AI exposes the illusion that speed equals intelligence. It reveals how often we confuse output with understanding, performance with learning, compliance with engagement. It forces us to confront a truth that teachers have always known: information alone does not produce meaning. Learning emerges through relationship, emotion, judgment, and design.

The chapters that follow trace this truth across time and practice. They begin before classrooms and technology existed, move through the emotional and moral architecture of learning, and arrive at the art of teaching in an AI-shaped world. Along the way, this book argues for teachers not as gatekeepers of information, but as guides – designers of environments where thinking, belonging, and courage can grow.

Technology, like AI, can generate answers. Teachers cultivate understanding.

That distinction has always mattered. Now, it is impossible to ignore.

INTRODUCTION

Teaching Has Always Been Design

Teaching has never been the delivery of information.

It has always been the design of environments.

Every classroom communicates something before a single word is spoken. The arrangement of desks signals whether collaboration is welcome. The timing of participation communicates whether speed is valued over thoughtfulness. The feedback patterns reveal whether revision is encouraged or correctness is rewarded. The first five minutes of class tell students what kind of thinkers they are expected to be.

We often talk about teaching as craft, as art, as instinct refined through experience. And it is. But before it is art, it is architecture.

Environments shape identity. Design shapes environments.

Students do not simply absorb content. They are formed by structure – by the invisible rules governing risk, participation, authority, attention, and revision. What they believe about themselves as learners is not accidental. It is constructed.

This book begins from a simple but urgent claim: Teaching must be intentionally designed for the kind of thinking we hope to cultivate.

When Answers Are Abundant

We now live in a moment when answers are increasingly easy to access.

Intelligent tools can draft essays, summarize research, generate examples, propose arguments, and simulate fluency in nearly any domain. Production has accelerated. Output appears polished. Efficiency has improved. But understanding has not automatically deepened. When answers become abundant, value shifts.

It shifts from production to judgment. From fluency to interpretation. From speed to discernment. From completion to formation.

The presence of powerful tools does not eliminate the need for teachers. It reveals whether our environments were ever designed for thinking in the first place.

If a classroom rewards quick responses, students learn that speed is intelligence. If it rewards polished products, students learn that performance is competence. If it rewards compliance, students learn that approval is the goal. But if a classroom requires explanation, revision, defense, transfer, and ethical positioning, students learn something different. They learn how to think.

The question is not whether technology can generate answers. The question is whether we have designed environments that cultivate what cannot be automated.

The Illusion of Knowing

For decades, education has confused production with understanding.

A well-written essay was assumed to reflect deep thinking. A correct answer was assumed to signal mastery. A fluent explanation was treated as evidence of comprehension. But fluency can mask fragility.

A student may reproduce a formula without understanding its structure.
They may repeat a definition without internalizing its meaning.
They may assemble a coherent paragraph without forming a coherent argument.

The acceleration of intelligent tools has not created this illusion. It has exposed it. When a system equates visible output with understanding, it becomes vulnerable to automation. Not because machines are superior thinkers, but because the system never required thinking deeply enough.

Designing teaching means refusing that illusion.

It means asking:

Where must students wrestle?
Where must they defend?
Where must they revise publicly?
Where must they transfer ideas across contexts?
Where must they confront ambiguity rather than eliminate it?

Understanding requires friction. Judgment requires tension. Confidence requires survivable risk. These do not emerge by accident. They are designed.

From Delivery to Architecture

If teaching is reduced to content delivery, then the most efficient delivery system will eventually outperform the human one. But teaching was never meant to be delivery. It is environmental architecture.

Teachers design:

Participation structures that determine who speaks and who remains silent.
Feedback loops that determine whether students fear error or use it.
Assessment systems that signal whether reasoning matters more than speed.
Norms that define whether disagreement is threat or growth.

These structures are not neutral. They cultivate identity.

A classroom can produce compliant students or courageous thinkers.
It can produce answer-seekers or question-askers.
It can produce dependence or discernment.

Design determines which.

This book argues that the central work of the educator is not protecting content from machines. It is designing environments that form judgment, resilience, and relational intelligence under conditions of acceleration.

Technology changes the landscape. It does not change the architecture of human formation.

The Educator as Designer

To design is to decide intentionally.

It is to ask:

What does this routine communicate?
What does this grading pattern reinforce?
What does this participation structure assume about intelligence?
What kind of risk is survivable here?
What kind of thinking is rewarded?

When design is unexamined, default patterns dominate — speed over depth, performance over understanding, efficiency over formation.

When design is intentional, different outcomes emerge — cognitive endurance, adaptive expertise, ethical judgment, disciplined originality.

Teaching is not a reaction to technological change. It is a profession grounded in shaping environments where human capacities are cultivated.

In an accelerated world, that work becomes more visible — and more necessary.

This book is not a defense against machines.

It is an invitation to design deliberately.

What This Book Is – and Is Not

This is not a technology integration manual. It is not a productivity system. It is not an argument for banning tools. It is a design framework for cultivating human judgment in environments shaped by intelligent systems.

Throughout these pages, we will explore:

- Relational architecture
- Cognitive tension and productive struggle
- Inquiry structures that require thinking
- Feedback systems that build identity
- Authority that cultivates influence rather than compliance
- Professional judgment under acceleration

The goal is not to compete with automation.

The goal is to design for what cannot be automated.

When environments are intentionally structured, students leave not merely with answers – but with agency. And that has always been the deeper purpose of teaching.

PART I — WHY TEACHING HAS ALWAYS BEEN ABOUT MORE THAN INFORMATION

Chapter 1: Teaching Before Chalk, Before Code

Long before classrooms had walls, teaching happened around fires. Knowledge was not written down; it lived in stories, gestures, repetition, and relationships. Elders taught not by lecturing, but by modeling - how to hunt, how to heal, how to survive winter, how to treat one another. Learning was communal, embodied, and deeply human.

As Barbara Rogoff (2003) explains, "In many Indigenous American communities, children are included in the mature activities of their community and learn by observing and pitching in to ongoing activities rather than by being segregated for instruction" (p. 326). Learning was not extracted from daily life; it was woven into it. Anthropological and educational scholarship consistently demonstrates that early learning systems across Indigenous cultures centered participation, observation, and interdependence. Teaching was not a separate event. It was living alongside one another with attention and purpose.

We see echoes of this truth in education even now, when we allow ourselves to slow down.

In an elementary classroom, a teacher tapes a large grid on the floor. Instead of writing multiplication problems on the board, he asks students to stand and form arrays with their bodies. Four rows of three. Three rows of four. Students rearrange themselves physically, noticing that while the formation changes, the total remains constant. They step forward and backward, count aloud, rotate positions. Laughter and minor confusion ripple through the room as they adjust.

The concept is not explained first; it is experienced.

When the class returns to their desks, the numbers on the page feel less abstract. Students no longer memorize that 3 × 4 equals 12; they have stood inside it. They have felt the structure shift while the quantity remained. The mathematics has moved from symbol to lived pattern.

Nothing about this lesson is technologically advanced. It is ancient. Learning through movement, observation, and participation is not innovative pedagogy; it is original pedagogy.

And yet, much of modern schooling does not resemble this.

The U.S. public education system began taking recognizable form in the mid-1800s. The common school movement sought to establish "a system of free public schools" aimed at promoting social stability and civic order (Kaestle, 1983, p. 98). In 1852, Massachusetts enacted "the first compulsory school attendance law in the United States" (Goldin, 1999, p. 2), formalizing schooling as a civic expectation. By the late nineteenth century, the common school had become "the principal institution for shaping republican character" (Cremin, 1957, p. 9), marking the consolidation of publicly funded, state-supported basic education nationwide.

These efforts, as we know, were deeply shaped by racial, gendered, and economic inequities. Still, it is notable that the system's original emphasis was not on learning itself, but on presence – who showed up and who was compelled to do so. Butts in seats.

Subsequent school laws, curricular mandates, and accountability systems have ebbed and flowed ever since, reflecting political priorities more than pedagogical realities. What these systems often fail to capture is teaching and learning as lived experiences.

Somewhere along the way, we began confusing teaching with delivery – as if learning were something that could simply be handed over intact.

When learning becomes measured primarily by presence, pacing guides, and performance indicators, embodiment disappears. Participation becomes compliance. Stillness becomes evidence of engagement. Coverage becomes mistaken for comprehension.

Artificial intelligence exposes that confusion.

When a machine can instantly generate explanations, essays, and solutions, it becomes painfully clear that access to information was never the teacher's greatest value. If it were, teaching would already be obsolete.

AI can calculate 3 × 4 in milliseconds. It cannot inhabit the lived experience of that structure. It can generate a paragraph explaining multiplication. It cannot notice when structure becomes visible to a student.

Research confirms what teachers have always sensed: memorization is not the same as understanding. "To develop competence in an area of inquiry, students must: (a) have a deep foundation of factual knowledge, (b) understand facts and ideas in the context of a conceptual framework, and (c) organize knowledge in ways that facilitate retrieval and application." (Bransford et al., 2000, p. 16). Science names what teachers practice daily. And as Darling-Hammond et al. (2020) remind us, "the brain is biologically wired to learn from social contexts and relationships" (p. 99).

Understanding does not emerge from information in isolation. It grows through interaction, feedback, emotion, and lived context.

Teaching is not obsolete because teaching has never been primarily about information.

What teachers do – and what they have always done – is help learners make sense of the world and their place within it. They notice when confusion hides behind compliance, when a student can produce the answer but cannot yet explain the idea. They attend to emotion, culture, history, and identity. They adjust – not because an algorithm instructs them to, but because they sense that something is not landing. As Hatano and Inagaki (1986) explain, "Adaptive experts are able not only to apply procedures efficiently but also to modify their knowledge and invent new procedures when faced with novel situations" (p. 266). Teaching lives in that flexibility – an expertise grounded not in routine procedure, but in responsive judgment.

Teaching did not begin as delivery. It began as participation.

And in a system that once prioritized attendance over understanding, and now risks prioritizing output over meaning, that origin story matters more than ever. This perspective becomes clearer when we examine how teaching unfolds across different educational contexts.

Throughout this book, examples span from elementary to university settings, reflecting different developmental stages and contexts. While the setting shifts, the underlying human dynamics remain constant. At any age, the tension between performance and understanding, fluency and ownership, efficiency and formation can persist. Across contexts, the craft of teaching becomes less procedural and more deeply rooted in judgment, relationship, and design. To help readers examine these dynamics within their own practice, each chapter concludes with reflection questions.

Chapter reflection questions invite readers to move from ideas to application. Some prompts support immediate analysis of classroom routines, while others encourage deeper examination of identity, assumptions, and long-term impact. This layered structure allows educators at different stages of practice to engage at an appropriate depth and revisit insights over time.

For educators seeking more structured application, *AI and the Art of Teaching Workbook* provides expanded tools and protocols that translate the book's principles into classroom design and practice.

Rather than prescribing change, the questions function as design tools that help educators diagnose, refine, and align daily decisions with their professional convictions. Positioned at the end of each chapter, they serve not as tasks to complete but as invitations to pause and interpret practice before acting.

In an era that rewards speed and immediate solutions, disciplined reflection protects professional judgment—ensuring that instructional decisions arise from intention rather than reaction, and that human formation remains central as systems, tools, and technologies compete for attention.

Reflection Questions

Chapter 1: Teaching Before Chalk, Before Code

1. If AI can generate explanations instantly, what uniquely human work remains at the center of your classroom?

2. What do your participation structures, grading patterns, and feedback loops quietly teach students about intelligence?

3. What would remain of your teaching if content disappeared but relationship remained?

4. What moments in your classroom cannot be quantified, yet feel essential to learning – and how do you protect them?

5. When technology accelerates information delivery, what slows you down intentionally – and why?

Chapter 2: Reading What Isn't Said

Teaching has never been just about delivering information.

Students may complete assignments correctly, answer questions accurately, or submit polished essays, and AI or standardized assessments could mark them as "on track." Yet a closer look reveals that correctness alone rarely tells the full story.

Some students avoid eye contact, shrink in discussions, or hesitate to ask questions, signaling fear, uncertainty, or prior experiences of ridicule. Skilled teachers notice these subtle cues – what is not being said – and respond with intentional strategies: small-group discourse, low-stakes rehearsal, think-pair-share, reflective conferences, and explicit validation of effort and perspective.

By attending to the unspoken barriers to participation, teachers create conditions where students feel safe to take risks, voice ideas, and wrestle with complexity. In these moments, learning deepens not because the content changed, but because the environment, the relational scaffolding, and the invitation to engage authentically shifted. This is the art of teaching.

Having been a K–12 teacher, having taught preservice teachers at university, and having evaluated teachers as a principal, I am confident that not all teachers enter the classroom with a refined art of teaching. Some struggle to develop it at all. Thankfully, many do – and many more are still in the process. As Feiman-Nemser (2001) reminds us, "Learning to teach is a long-term process that extends well beyond initial preparation" (p. 1013). Expertise is not conferred at the completion of a teacher preparation program or at university graduation; it is cultivated through reflection, mentorship, and lived experience. New teachers enter the profession not as finished experts, but as vital contributors whose questions, energy, and emerging insights renew and strengthen the field itself. The art of teaching takes time, unfolding differently in each teacher who chooses to stay in the work.

The art of teaching is deeply reminiscent of storytelling. Stories organize information, evoke emotion, and anchor meaning in memory. As Immordino-Yang and colleagues (2019) explain, "Emotion and cognition are deeply integrated in the brain and in behavior, and emotion plays a critical role in learning" (p. 185). Narrative engagement activates networks associated with reflection and meaning-making, enhancing comprehension and retention. There is much to be learned from stories – not only what they communicate, but how they are delivered.

Storytelling, after all, is not passive recitation; it is performance. In *The Big Bang Theory*, Season 4, Episode 14 ("The Thespian Catalyst"), Dr. Amy Farrah Fowler quips that teaching is, in its essence, a performance art. The line is humorous – and one I find profoundly accurate. Performance does not imply artificiality; it implies attunement. Teachers read the room. They pace and pause. They emphasize and improvise. They cultivate engagement, sustain attention, and invite understanding. They wear spirit-day outfits, tell stories drawn from lived experience, and connect content to students' lives in ways that make it matter.

When I ask students to reflect on their most powerful educational experiences, they seldom point to silent note-taking. Instead, they recall movement, debate, and moments of intellectual risk. They remember being required to defend a position, to respond in real time, to think beyond prepared answers. Interactive moments endure because they require presence; they ask something of the learner. It is often in these moment of structured participation that teaching moves beyond lecture altogether.

Rather than explaining the branches of government through slides, a teacher assigns roles: legislators drafting bills, journalists questioning intent, citizens advocating for or against, judges weighing constitutionality.

Rather than describing food webs and trophic levels, each student becomes an organism – producer, herbivore, predator, decomposer, invasive species – connected by strands of yarn that trace the flow of energy through a living system.

Instead of lecturing on theme and character motivation, the classroom becomes a courtroom. The protagonist stands accused – of betrayal, cowardice, injustice, moral failure – while students serve as attorneys, witnesses, and jurors citing textual evidence.

Rather than explaining slope through formula alone, the teacher tapes a coordinate plane on the floor. Students become ordered pairs. One group forms a line. Another adjusts to make it steeper.

In each case, expertise is not abandoned; it is staged. The teacher designs the experience, anticipates friction, and orchestrates interaction. Students are not passive recipients of explanation. They are participants in meaning-making.

Lecture has its place when clarity, synthesis, or foundational framing is needed – but it cannot substitute for the deeper understanding that emerges through participation. When learning is experienced – argued, modeled, tested, embodied – understanding becomes durable. Teaching, again, is performance art – not because the teacher dominates the spotlight, but because the teacher knows when to step back and let the thinking take center stage.

Teaching, at its most artful, is not the transfer of prepared words. It is the orchestration of experience. This is true in and beyond school. Some of the greatest teachers in our lives never stood at a whiteboard. One of mine was my friend Brian. He introduced me to the song *Wasn't That a Party* by The Irish Rovers, was quick with his wit, and generous with his time. In my 30s and 40s, he called me "kiddo," especially when telling stories. Though only nine years older than I was, Brian came from a different time. I have come to realize how deeply his phrases and stories shaped the way I see the world.

After his passing in 2026, his children compiled a list of his memorable sayings – many of which reveal how humor, metaphor, and exaggeration function as teaching tools: "Potholes—takes a good driver to hit 'em all." "What a soup sandwich." "Couldn't hit water if he fell out of a boat."

Brian was never at a loss for descriptive language. Brian did not correct; he reframed. He did not lecture; he narrated. He did not instruct; he invited. While I miss my friend deeply, I hold tightly to the wisdom and humor he shared. He showed up. He paid attention. He made meaning memorable. There is a lot that can be learned from showing up and paying attention.

Illustration in the Classroom: Elementary Level

Elementary

Students are reading about a moment of historical injustice.

The text describes unfair laws that affected a group of people in the past. The teacher pauses and asks, "Why do you think this law was unfair?"

Several hands rise quickly. Others remain still. One student keeps her eyes on her desk. She understands the story. She can retell the events accurately. But discussing injustice feels different from answering comprehension questions. It feels personal. She worries about saying something wrong – about misunderstanding the history, about offending classmates, about exposing her own family's experiences. Her silence is not confusion. It is caution.

An AI system might interpret her quiet as disengagement. Her teacher notices something else.

Rather than calling on volunteers randomly, her teacher shifts the structure. Students first write privately in response journals. Then they turn to a partner and share one sentence they feel comfortable discussing. Sentence starters are offered: "I think this was unfair because..." and "This reminds me of..." The teacher models vulnerability, acknowledging that conversations about fairness can feel complicated and that learning often involves sitting with discomfort.

The teacher establishes clear norms: Listen fully. Disagree with ideas, not people. Assume good intentions. Ask questions before making conclusions.

The room settles.

When discussion begins again, the student speaks quietly to her partner. Later, during whole-class conversation, she raises her hand – not to deliver a perfect answer, but to offer a thoughtful observation about how laws can protect some people while hurting others. Her voice is tentative, but steady.

What changed was not the content. It was the architecture of safety around it.

Elementary

Civic reasoning requires more than recalling facts. It requires moral reflection, perspective-taking, and the courage to examine complexity. Without emotional safety, students default to silence or surface-level agreement. With emotional safety, they begin to analyze systems, question fairness, and consider multiple viewpoints.

The teacher did not simplify the history. He scaffolded the conversation.

Artificial intelligence can summarize historical events and define injustice with precision. It cannot sense when a student's silence reflects vulnerability. It cannot create the relational conditions that allow moral reasoning to emerge.

In this classroom, the teacher is not only teaching about the past. He is teaching students how to think ethically in community – how to examine injustice without fear, and how to use their voices responsibly.

Understanding deepens not because the facts change, but because students feel safe enough to wrestle with them.

Illustration in the Classroom: Middle School Level

A student answers quickly and often. He jokes during instruction.

He calls out responses without raising his hand. On the surface, he appears distracted – perhaps even defiant. A behavior tracking system flags him for disruption. The pattern is clear: frequent interruptions, side comments, off-task behavior during whole-group instruction. But the teacher notices something else.

Humor surfaces most often when new vocabulary is introduced or when abstract concepts require sustained reading. During silent work time, he stares at the page longer than his peers, then cracks a joke. When asked to read aloud, the humor sharpens. The timing is not random. It is protective.

Through quiet observation and private conversation, the teacher learns that reading has never felt easy for him. Complex texts feel like exposure. Humor keeps the spotlight from lingering too long. Laughter preserves status when confidence wavers. The issue is not disrespect. It is vulnerability.

Instruction shifts subtly and intentionally. Key vocabulary is previewed before whole-class reading. Visual supports accompany new concepts. Students rehearse responses with partners before sharing publicly. Short, structured reading segments replace long, uninterrupted passages. During discussion, the teacher names effort rather than speed. "I noticed you stayed with that paragraph even when it was dense." Humor is gently redirected.

Gradually, jokes decrease because they are no longer necessary. When comprehension strengthens, the need for deflection diminishes. Participation becomes steadier, less performative.

What changed was not the curriculum. It was the interpretation of behavior. AI cannot detect when humor is armor or when laughter protects dignity. Understanding is often hidden beneath performance. And misbehavior can be a form of communication. Teaching, at its most perceptive, requires noticing the difference.

Illustration in the Classroom: High School Level

A student demonstrates extraordinary technical skill.

Her shading is precise. Her proportions are accurate. When asked to replicate a still-life image, she produces near-photographic accuracy.

But when invited to reinterpret the subject – to distort perspective, shift medium, or communicate personal meaning – she resists. "Just tell me what you want," she says.

Her technique suggests confidence. Her avoidance suggests something else.

The teacher notices that risk, not skill, is the barrier. During critique sessions, he reframes evaluation. Instead of praising realism alone, he asks about intention. "What were you trying to say?" "What would happen if you exaggerated this?" He shares examples of artists whose early work was technically strong but conceptually restrained. He models vulnerability by sharing unfinished pieces of his own.

Gradually, the student begins to experiment. A still-life becomes symbolic. Shadows lengthen with intention. Color shifts to evoke mood rather than accuracy. Her technical fluency does not disappear – it deepens. It begins to carry voice.

Artificial intelligence can evaluate outputs. It can assess symmetry, detect design inconsistencies, and measure technical precision. But it cannot perceive the subtle pause before creative risk. It cannot sense the tightening posture that signals fear of exposure. It cannot distinguish between mastery and self-protection.

Understanding is often present long before it is visible. And insecurity is often hidden behind competence.

Teaching, at its most artful, requires seeing beyond what students produce – and attending carefully to what they hesitate to attempt.

What changed these learning trajectories was not new content, but relational attention – the teacher's willingness to notice what lay beneath the behavior and respond with humanity rather than correction.

As Hammond (2015) reminds us, "Culture, it turns out, does matter for learning." Students' communication styles, definitions of respect, comfort with public discourse, and responses to authority are shaped by cultural and social context long before they enter the classroom. When educators misread those cues, difference can be mistaken for deficit.

Relational attention is not sentimentality; it is cognitive alignment. It is not warmth alone; it is the disciplined noticing of how emotion and identity shape cognitive availability. Rational attention asks what the student is attending to — and what they are protecting. Cognitive alignment occurs when instructional design meets readiness, allowing challenge without threat. When alignment is absent, students defend. When it is present, they engage.

Classroom behavior is best understood as communication. "Problem behavior serves a function for the student and is related to the context in which it occurs" (Sugai et al., 2000, p. 132). It signals unmet needs, inaccessible instruction, or environments that do not yet feel safe. Artificial intelligence may detect patterns, but it cannot interpret meaning.

Understanding is often present long before it is visible, and teaching requires patience, curiosity, and trust to bring it forward. My friend Brian taught in much the same way. His stories were not about instruction but about invitation – an invitation to see differently, to remember through metaphor, to feel safe enough to laugh at confusion. Like the teachers in these examples, Brian did not rush to fix or explain; he stayed, observed, and offered language that made meaning stick.

The learning that lasts is rarely the information exchanged; it is the sense of being known. Across ages and contexts, that remains the quiet work of real teaching.

Reflection Questions

Chapter 2: Reading What Isn't Said

1. What behaviors in your classroom might be communication rather than compliance – or defiance?

2. What assumptions shape how you interpret student behavior – and how might those assumptions limit what you see?

3. When was the last time you adjusted instruction because of something subtle – a facial expression, posture shift, or tone?

4. Where might AI-generated data oversimplify what is actually happening beneath student performance?

5. How does the way you interpret behavior influence the identities students begin to construct about themselves?

Chapter 3: Confidence is Curriculum

Research confirms what teachers witness daily: belief precedes performance. "Students' beliefs about their intelligence have a powerful impact on their learning and achievement" (Dweck, 2006, p. 6), and students with stronger self-efficacy demonstrate "greater persistence over time" (Schunk & DiBenedetto, 2020, p. 4).

These findings are not abstract psychological claims. They describe the quiet dynamics teachers observe every day. Students who believe effort will matter attempt difficult tasks more readily. They revise work that initially fails. They ask questions when confusion appears. In contrast, students who doubt their capacity often withdraw before learning has a chance to unfold. The difference rarely lies in raw ability. It lies in whether struggle is interpreted as evidence of growth or as proof of limitation.

In other words, what students believe about themselves as learners frequently shapes what they become capable of achieving. Performance matters – but confidence determines whether students persist long enough to grow.

This relationship explains why teaching is never only about content. When classrooms communicate that mistakes are survivable, effort is meaningful, and revision is expected, students remain present long enough for understanding to deepen. When learning environments communicate the opposite – when correctness is prized above growth – students begin protecting themselves from exposure. They retreat into silence, speed, or superficial compliance. In either case, beliefs quietly shape the trajectory of learning.

If teaching is relational formation rather than information delivery, then confidence is not a byproduct of instruction; it is part of the curriculum itself.

Teachers cultivate it through small, often unnoticed choices: the way they respond to uncertainty, the patience they extend during struggle, the questions they ask when answers falter. These signals accumulate. Over time, students begin to interpret themselves differently—not simply as performers of tasks, but as thinkers capable of growth.

I did not understand this fully on my first day as a teacher.

On that morning, I stood alone in an empty classroom, staring at rows of desks waiting to be filled. The room was quiet, but my thoughts were not. I felt like an imposter rehearsing the phrase I had long relied on—*fake it till you make it.* I carried life experience with me, but none of it felt equal to the authority I imagined the role required. I had mistaken the position for certainty. Instead, I felt exposed.

My husband had persuaded me to attend the interview that led to this job. "You've got this," he said, framing it as practice and seeing more in me than I did. I believed I had a safer, more predictable offer waiting at a comprehensive school. I did not fully understand what I was stepping into when I interviewed for a position in a residential incarcerated school. But something in that interview unsettled me in the best way. I left wanting the job—deeply, unexpectedly—and when the offer came, I said yes before the sentence on the other end of the line had even finished forming.

Now, facing those empty desks and anticipating the arrival of students whose lives I could not yet imagine, doubt crept in. I questioned my preparation. I questioned my knowledge. I questioned whether I belonged in that room at all. When the students finally walked in, I remember feeling like the shortest person there—not only in height, but in certainty. *Why was I here?*

And then my mother's voice surfaced, steady and familiar:
"Yesterday is history, tomorrow is a mystery, today is a gift; that's why it's called the present."

Confidence did not arrive as certainty that morning. It arrived as decision. I did not know everything. I was not fully formed. But I could be present. I could listen. I could learn alongside my students. And that, I would come to understand, was enough to begin.

Illustration in the Classroom: Elementary Level

In one classroom, confidence wears two disguises.

One student answers quickly and often. He blurts out responses before others have time to think. When corrected, he shrugs and smiles. Sometimes he jokes. Sometimes he insists he already knows the answer. The data is clean: impulsivity, interruption, limited wait time. A behavior tool can name it instantly.

But the pattern has texture data cannot capture.

The blurting happens most often when the math is new – when the concept has not yet settled into certainty. During independent work, his pencil pauses midair. He scans the room before writing. In partner talk, he defers quickly if he senses he might be wrong. The first answers are not always accurate. They are fast. They are protective.

Across the room, another student presses her pencil hard against the paper, then stops. Sometimes she crumples the page. Sometimes she leaves sentences unfinished. Over time, she has begun telling herself she is "not good at writing." Her silence is not speed. It is retreat.

Two students. Opposite behaviors. The same architecture beneath them.

One protects competence through urgency. One protects identity through avoidance.

The teacher does not treat these as personality traits. She treats them as design problems.

In math, she builds in think-time before any voice enters the room. Everyone writes before anyone speaks. Speed loses its advantage. She models revision publicly without drama – "I almost rushed that too" – and treats recalibration as normal thinking, not correction. In conferences, she names the moment he pauses before blurting as evidence of growth. She praises not quickness, but staying.

Elementary

In writing, she restores the relationship to effort. She names small victories with specificity: a complete sentence, a stronger verb, a brave attempt at dialogue. She drafts aloud, crossing out lines without apology. "Watch how I fix this." Struggle becomes visible—and therefore safe.

Peers become part of the architecture. Ideas are rehearsed orally before being written. Students borrow phrases as bridges into their own. Feedback begins with meaning "What are you trying to say?" – before mechanics.

The shift is not immediate. It is cumulative.

His hand still rises – but later. Her pages are no longer crumpled – but revised. He begins checking reasoning before offering it. She begins lengthening sentences before abandoning them.

Confidence here is not bravado and it is not praise. It is infrastructure. It is the slow construction of conditions where being unfinished is not dangerous. Neither student needed more intelligence. They needed different architecture.

Confidence does not arrive all at once. It is built – through structures that make risk survivable.

Illustration in the Classroom: Middle School Level

Confidence is negotiated over time and through experiences.

One student listens closely, nods when others speak, and keeps her eyes on her notes. When called upon, her responses are brief and carefully measured. Silence, here, is not emptiness. It is calculation.

Each sentence carries risk: the right word, the right tense, the right pronunciation. A pause can feel like exposure. A misstep can become a story classmates repeat. Over time, caution becomes identity: I am the careful one. I am the one who should not draw attention.

Another student completes his lab work meticulously. Measurements are exact. Tables are tidy. Procedures are followed precisely. But when asked to interpret results – to explain why bubbles formed more rapidly in one trial than another -his eyes drift toward peers, searching for confirmation. When uncertainty enters, confidence retreats.

Two students. Different subjects. The same tension.

Participation feels public. Interpretation feels personal.

The teacher does not treat confidence as temperament. She treats it as architecture.

In discussion, she shifts how ideas enter the room. Students rehearse in pairs before speaking publicly. Sentence starters reduce cognitive load so energy can be invested in meaning rather than form. Questions are sequenced: What do you notice? What might it suggest? What evidence supports that claim?

Discussion roles redistribute power – summarizer, connector, challenger, clarifier – so interpretation becomes collaborative rather than performative. Speaking is no longer a test of fluency. It is a shared construction of meaning.

In science, interpretation becomes group practice rather than solo exposure. Students must generate two competing explanations before choosing one.

They identify evidence that supports each claim and name what remains uncertain. When an explanation collapses under scrutiny, it is treated as progress: a discarded theory is still thinking.

The room begins to feel different – not quieter, not easier – but safer.

She experiments with vocabulary without apology. He offers predictions before seeking reassurance.

Her sentences lengthen. His reasoning deepens.

Neither student became suddenly bold. The structure changed first.

Confidence at this age is not loudness. It is agency plus belonging. It is the freedom to think publicly without demanding immediate certainty. It is the ability to revise mid-sentence without interpreting that revision as failure.

Middle school learners are exquisitely attuned to peer perception. The architecture must account for that. When participation is treated as personality, the quiet grow quieter and the cautious grow more rigid. When participation is designed intentionally, competence becomes distributable.

Confidence here is not praise. It is patterned experience.

It is the repeated encounter with uncertainty under conditions that do not threaten identity.

Illustration in the Classroom: High School Level

In high school, confidence is rarely about participation alone.

It becomes entangled with identity.

In one lab, a student mislabels slides during a microscopy assessment. The mistake is small. The score is not. After that day, her posture changes. Precision becomes ritual. She double-checks every step, rereads directions twice, moves more slowly than before. When asked to design an experiment, she defers. "I don't want to mess it up again," she says quietly.

Across the hall, in a literature seminar, another student submits essays that are structurally strong and analytically sound. Her thesis is clear. Quotations are integrated smoothly. But during discussion, something shifts. She begins to explain an interpretation, then pauses. A sentence trails off. She restarts and abandons the thought again. The presence of peers turns analysis into exposure. A partially formed idea feels public. A misreading feels permanent.

Two classrooms. Different disciplines. The same architecture beneath them.

In both spaces, one visible error has begun to function as identity.

The science student no longer fears content; she fears being seen as careless. The literature student does not fear text; she fears being seen as wrong.

The teacher does not respond with reassurance alone. She redesigns what the classroom publicly rewards.

In the lab, reflections now ask: "What surprised you?" "What would you revise?" "What assumption did your data challenge?"

Demonstrations include flawed trials – not as cautionary tales, but as the expected shape of scientific work. Rubrics shift. Reasoning and adaptive thinking carry weight alongside correct outcomes. During debriefs, the teacher narrates his own uncertainty as part of method. "At first I thought this meant... but the data resisted that interpretation."

High School

Interpretation is rehearsed before it is declared. Students test readings in small groups before whole-class discussion. Counterarguments are anticipated, not feared. The teacher models revision mid-thought: "I'm reconsidering this line. Let's look again." Disagreement becomes inquiry rather than correction.

The room changes – not in volume, but in posture.

She begins proposing experimental variations again. She traces her reasoning aloud before it is perfect. She offers a second interpretation of a metaphor and holds two possibilities in tension. She revises mid-sentence and is not punished for it.

The missed slide becomes information rather than indictment. The abandoned sentence becomes thinking rather than failure.

Confidence at this age is not loudness. It is endurance. It is the willingness to think publicly without demanding immediate certainty. It is the capacity to revise without collapsing.

Adolescents are forming durable narratives about competence. One visible error can calcify into "I am not a scientist" or "I am not insightful." If classrooms equate correctness with credibility, confidence narrows quickly. If classrooms normalize revision as method, competence expands.

Confidence here is not affirmation. It is disciplined exposure under protective structure.

It is the repeated experience of recovering from error in public – and discovering that identity remains intact.

Students benefit from relational scaffolding, even when struggles are not tied to academic skill. This scaffolding can look like teachers who emphasize effort, strategy, and persistence rather than compliance alone. "We have to get the brain ready for rigor" (Hammond, 2015).

The consistent message that growth matters more than perfection does more than protect self-esteem; it cultivates agency. As Bandura (2001) reminds us, "Among the mechanisms of human agency, none is more central or pervasive than people's beliefs in their efficacy to exercise control over their own functioning" (p. 10). Agency begins with the quiet conviction that we can influence our own becoming.

When students come to believe they can grow, they begin to act differently – choosing challenge over avoidance, effort over retreat. Schunk, Pintrich, and Meece (2008) note that "Students with strong self-efficacy are more likely to engage in tasks, persist when difficulties arise, and achieve at higher levels" (p. 107). Over time, these beliefs shape participation, performance, and identity across subjects and social contexts. Growth, consistently reinforced, becomes not just a strategy but a way of seeing oneself.

Growth matters more than perfection.

Across contexts, across classrooms, confidence functions as instructional infrastructure. It shapes participation, risk-taking, resilience, and sustained engagement. Without it, students may comply – but they rarely flourish. It is the quiet architecture beneath learning, the unseen framework that allows courage to rise and effort to endure.

Through consistent relational support, students develop confidence and resilience. "Self-efficacy influences students' effort, persistence, and resilience" (Schunk et al., 2008, p. 124), and socially and emotionally competent students are "more likely to engage in positive lifelong behaviors" (Zins et al., 2004, p. 7). Emotional scaffolding in classrooms does more than support the moment; it lays the foundation for how students approach learning long after the lesson ends.

What begins as classroom support becomes a way of being. As such, teaching is responsive in ways that artificial systems often are not.

This responsiveness is why students remember their teachers long after specific content knowledge has faded. They recall the moments when a teacher made them feel capable, seen, or safe enough to take risks. While teachers can impart knowledge that provides immediate information, it is the cultivation of confidence, curiosity, and problem-solving habits that equips students with enduring understanding. "The aim is to teach students to become their own teachers" (Hattie, 2009, p.22). By fostering these qualities, educators ensure that their impact lasts far beyond the classroom.

Artificial intelligence can generate explanations. It cannot cultivate trust. It can optimize efficiency. It cannot read a room. It can predict patterns. It cannot understand meaning. What AI cannot perceive, teachers must.

Teachers notice when confusion is masked by compliance. They recognize when a student understands the answer but not the idea. They respond to emotion, context, culture, and history. They adjust – not because an algorithm tells them to, but because they sense something isn't landing.

Teaching has always been relational labor disguised as intellectual work; AI simply makes that truth visible. This relational dimension of teaching is not sentimental; it is methodological.

In the book *Street Data*, Safir and Watson (2021) reminds us that "Street data are qualitative data that are at eye level and on the ground, offering a more complete story about students' lived experiences" (p. 5). When teachers listen carefully to those stories – to students' voices, histories, and contexts – they move beyond surface performance into deeper understanding. Safir and Watson argue that "We must listen to students' voices and experiences in order to create conditions for deeper learning" (p. 32). This responsiveness fosters trust, invites open dialogue, and reveals what compliance can conceal. It strengthens the relational dynamics at the heart of effective teaching.

Research on relational pedagogy confirms what teachers have long practiced: "The quality of teacher–student interactions is one of the most important determinants of student engagement and achievement" (Pianta et al., 2012, p. 369). Students at risk of school failure, in particular, "benefit most from emotionally supportive teacher–student interactions" (p. 373). Teaching has always been relational. The subtleties of noticing, supporting, and guiding students cannot be replicated by technology alone.

Rather than asking how teaching must change to keep up with artificial intelligence, we might ask: What does AI reveal about what teaching was never meant to be?

In addressing this question, we recognize the unique contributions of educators in the learning process. Teachers are the connective tissue of learning. They translate content into meaning, notice what is not being said, and respond to the emotional and cognitive realities students bring each day. In an age where information is abundant and answers are instantaneous, teachers remain essential – not because they know the most, but because they cultivate judgment, guide thinking, and create the conditions in which learning can truly occur.

The art of teaching is relational, adaptive, and deeply human work. Artificial intelligence may change how instruction looks, but it cannot replace the teacher's role in helping students make sense of the world and themselves within it.

Invitation

This book is written with teachers in mind. Every argument, example, and question assumes the professional presence of educators who are already engaged in the daily work of designing learning, responding to students, and navigating complexity. It speaks directly to the thinking, judgment, and humanity teachers bring to their practice, honoring their expertise rather than simplifying the work they do.

Reflection questions throughout the chapters invite readers to examine their own instructional decisions, while the companion volume, *AI and the Art of Teaching Workbook*, offers structured tools and protocols for translating these ideas into classroom design and practice.

Additional invitations to engage with this work are offered for preservice teachers, administrators, and homeschool parents, each of whom stands in a different relationship to the act of teaching – preparing to enter it, shaping the conditions that sustain it, or enacting it outside formal systems.

Each group plays a distinct and necessary role in the formation of learners, whether by stepping into the work, designing the environments that support it, or carrying it forward in spaces beyond traditional classrooms.

Future volumes in this series will explore these perspectives more fully, with books designed specifically for preservice teachers, educational leaders, and families who support learning beyond traditional school settings.

Invitations for Preservice Teachers

Preservice teachers stand at the threshold of this profession.

Preparation is less about mastering tools and more about cultivating professional judgment, reflective habits, and a deep understanding of learning as a human process. In an era where AI can generate plans, explanations, and materials with ease, preservice teachers must learn to distinguish between instructional efficiency and educational purpose. Becoming a teacher is the gradual cultivation of discernment, presence, and ethical responsibilities – capacities no technology can supply.

As you begin this work, remember that teaching is far older than schools, standards, or technology. It did not originate in lesson plans or assessment systems; it began in relationships. Learning has always unfolded through attention, modeling, correction, encouragement, and presence.

As AI accelerates the production of explanations and instructional materials, your responsibility becomes clearer. Your task is not to move faster than technology but to notice who your students are becoming in your presence. Discernment is not about cognitive speed – it is about recognizing identity signals, relational shifts, and moments that will never appear in a dataset. AI may reshape classrooms, but it cannot replace the human work of remaining present as learners form their understanding of themselves and the world.

You are invited to consider: What kind of teacher are you becoming beneath the strategies you are learning? When efficiency tempts you, what anchors you to purpose? How will you cultivate presence in a profession that often rewards productivity?

The question is not whether instruction can be produced efficiently. It is whether you can help students become confident, capable meaning-makers. That work begins – and endures – with you.

Invitations for Administrators

Administrators shape conditions under which teaching is possible.

The systems, expectations, and structures you design either support or constrain meaningful learning. In an AI-influenced educational landscape, leadership decisions about assessment, policy, professional trust, and instructional priorities determine whether technology enhances thinking or undermines it. Through these choices, administrators signal what schools value: depth or speed, understanding or compliance, trust or surveillance.

Schooling and teaching are not synonymous. Schedules, policies, and curricula organize learning, but they do not generate it. Learning emerges through human interaction – in the exchanges between teacher and student that shape identity, confidence, and intellectual courage.

Leadership in this era requires more than technological fluency; it requires philosophical clarity. Protect the relational spaces where formation occurs. Efficiency is measurable; formation is not. Artificial intelligence will amplify whatever priorities already exist in a system. If efficiency dominates, AI accelerates fragmentation. If relational teaching is protected, AI can serve as a supportive tool rather than a destabilizing force.

You are invited to consider: What is your system quietly teaching about what matters most? Do your structures privilege speed over depth, or formation over performance? Where might relational work be assumed rather than intentionally protected?

To lead well now is to safeguard the relational core of education – creating conditions where teachers are trusted as professionals, students are recognized as developing humans rather than data points, and innovation strengthens understanding rather than replacing it.

Invitations for Homeschool Parents

You occupy a unique and powerful teaching role.

Homeschooling blends instruction, relationship, and lived experience, often allowing curiosity, identity, and pacing to be honored in ways formal systems struggle to sustain. The most important learning in your home may never be the fastest or most visible.

In an era shaped by AI, homeschool parents help students interpret struggle, authority, success, and responsibility while developing habits of self-directed learning. You model how tools are used thoughtfully rather than reflexively and how learning connects to daily life rather than remaining compartmentalized.

Homeschooling reconnects learning to its oldest foundations: shared experience, observation, conversation, and trust. Long before institutions formalized education, learning unfolded beside adults through questioning, modeling, and participation. In many ways, homeschooling carries forward that original rhythm.

Meaningful education depends on intention, attentiveness, and relational presence. Technology may expand access, but your consistency, curiosity, and engagement with your student's questions give learning its depth.

You are invited to consider: What rhythms of learning define your home? Where is curiosity protected from haste? How do your daily choices communicate that thinking is woven into living, not isolated from it?

Education has never required institutions to be meaningful. It has always required presence.

Reflection Questions

Chapter 3: Confidence is Curriculum

1. If confidence functions as instructional infrastructure, how intentionally are you designing it?

2. What narratives about intelligence are students internalizing in your classroom?

3. When have you witnessed confidence grow before performance improved?

4. Where in your classroom do students experience success after struggle – not just correction after error?

5. How do you ensure that encouragement strengthens agency rather than dependence?

Chapter 4: The Illusion of Knowing

There is a moment every teacher recognizes. A student finishes quickly. The answers are correct. A hand shoots up with confidence. And yet, when asked to explain why the answer works, certainty falters. Words trail off. The understanding, it turns out, is thinner than it first appeared. It is a humbling moment when that recognition turns inward.

Sometimes the illusion is not in the student's answer, but in the teacher's assumption. I once assumed that certain aspects of adulthood were universally understood. During a financial literacy lesson, I was explaining how employer-sponsored health insurance works - premiums deducted from paychecks, deductibles, co-pays, coverage networks. I spoke as if I were clarifying complexity. I believed I was refining prior knowledge.

Halfway through the lesson, a student raised his hand and asked, without irony, "Wait... you have to pay for insurance even if you don't go to the doctor?" Another followed, "My family doesn't pay. We just go." Heads nodded in agreement around the room.

In that moment, I realized I had built the lesson on an assumption of shared experience. I had treated insurance as a universal system rather than a differentiated one. Some students had only known public assistance programs. Others had never seen parents navigate bills and networks with quiet stress. What felt obvious to me was not neutral knowledge; it was contextual knowledge.

They could copy the vocabulary. They could complete the worksheet. But the structure itself – why someone pays monthly for something they might not use – was unfamiliar terrain. I had mistaken exposure for understanding.

That day altered how I framed instruction. I began asking not only, "Do you understand this?" but "What experiences shape how this makes sense to you?" I stopped assuming that fluency meant familiarity. I learned to listen for what was not being said – the silent gaps where lived experience diverged from curricular assumption.

Understanding is fragile when we build on foundations we have not first uncovered.

These moments are familiar across grade levels. They reflect a fundamental distinction in learning: knowing the answer versus understanding the concept. As Bransford and colleagues (2000) observe, "Students often acquire disconnected sets of facts that they can recite but do not understand" (p. 16). Correctness, in these cases, is not comprehension; it is recall without structure.

Teachers constantly balance correctness and comprehension. While performance matters, understanding is far more durable and transferable. As the National Research Council (2012) explains, "Transfer is facilitated when learners understand the underlying principles that apply across situations" (p. 92). It is precisely this difference that technology now places under a brighter light.

This tension existed long before artificial intelligence, but AI makes it harder to ignore. It produces fluent explanations, polished reasoning, and correct answers instantly. To the untrained eye, such output can appear indistinguishable from understanding. Yet, as Perkins (1992) reminds us, "Knowledge that is merely inert does not help people think and act in the world" (p. 8). Knowing that something is correct is not the same as knowing why it matters.

We must be careful what we prioritize, the output or the understanding.

For generations, schools have quietly rewarded performance over comprehension. Memorization has passed as mastery. Speed has stood in for insight. Students have learned how to look like they understand, even when they do not. Teachers learn and know the difference; systems often do not.

There is a particular kind of confidence that resembles competence but is not anchored in understanding. It moves quickly, produces correct outputs, and rarely pauses to reflect. My friend Brian had a phrase for this kind of momentum: "Potholes – takes a good driver to hit 'em all." He said it jokingly, but the insight is sharp. Speed does not protect us from gaps; it often guarantees we encounter them without noticing.

In education, fluency can mask fragility. When students – or systems – prioritize momentum over meaning, misunderstandings are encountered faster, not resolved. AI intensifies this illusion by generating polished responses that feel authoritative, reinforcing the false belief that smoothness equals mastery. Teaching, then, is not about helping learners move faster; it is about helping them slow down long enough to notice where the road is uneven – and why that matters.

The pressure to move quickly does not originate with students alone; it is embedded in how schools are evaluated and funded.

Educational systems are often funded based on performance indicators rather than learning itself, and performance-based assessments continue to dominate curriculum design. "When important educational decisions are based on test results, those tests inevitably influence instruction" (Popham, 2001, p.44). As such, it is worth noting that "Accountability systems focused narrowly on standardized test scores can distort instructional practice" (Darling-Hammond et al., 2020, p.112). What we choose to measure inevitably shapes what we choose to nurture.

Funding structures, policy decisions, and accountability measures quietly shape these inequities. As Darling-Hammond (2010) observes, "Students in high-poverty schools are more likely to experience a narrowed curriculum focused heavily on basic skills and test preparation" (p. 40). When survival metrics dominate, creativity becomes harder to justify. In this way, educational systems reward visible performance more readily than durable understanding.

Innovative, creativity-focused programs remain less common and can be met with skepticism because their metrics are less traditional. In low socioeconomic areas, such programming may be limited, while higher socio-economic areas may enjoy greater access and flexibility. These disparities are not incidental. As the OECD (2025) observed, "advantaged students scored close to one standard deviation higher than disadvantaged students in the PISA 2022 creative thinking test," underscoring the quiet link between opportunity and creativity.

Funding structures and accountability systems amplify these patterns, rewarding what is immediately visible while leaving deeper formation unevenly distributed. What can be counted is often what is funded. Yet beneath these systems lies a simpler truth: teaching requires students. To cultivate critical thinking, animate knowledge, and design experiences that endure, teachers need engaged learners – not merely bodies present, but minds invested.

In the United States, public school positions are sustained through funding models tethered to metrics that often feel distant from the complexity of real learning. Enrollment numbers. Attendance counts. Butts in seats. These measures organize budgets and staffing decisions, even when they reveal little about curiosity, courage, or conceptual depth. Within this imperfect structure, another reality persists: to teach in schools, one must be employed. Credentials must be maintained. Professional standards met. Evaluations completed. Compliance fulfilled. Employment forms the framework within which public school teaching occurs. And so, however limited they may be, the metrics matter – even as they fall short of capturing what matters most.

For educators who want to make a lasting impact, understanding these realities is essential. Teachers who aim merely to 'skate by' – meaning they rely on recycled lesson plans without examining their effectiveness, prioritize coverage over comprehension, avoid difficult conversations about student misunderstanding, depend on worksheets or automated tools to generate surface-level engagement, and measure success by quiet compliance rather than visible thinking – may survive briefly, but both student learning and career longevity suffer.

Skating by in this context is not dramatic failure; it is subtle disengagement from the intellectual and relational demands of the profession. It is teaching on autopilot rather than with intention. This is where drift must give way to intention. If you are skating by, begin the quiet decision to turn off autopilot.

Artificial intelligence accelerates this exposure. When tools can generate materials, explanations, and even feedback instantly, the difference between thoughtful design and mechanical delivery becomes more visible. Systems increasingly reveal gaps in practice, making it harder for underprepared teachers to hide behind routine.

Remaining reflective, intentional, and engaged is no longer aspirational; it is necessary.

Illustration in the Classroom: Elementary Level

The question interrupts the neatness of the worksheet.

"Why do roots grow downward?"

Students glance back at their pages. One student – whose diagram is already complete – hesitates. She has labeled every part correctly. The vocabulary is secure. The arrows are precise. But explanation requires more than recognition. When asked how sunlight affects growth, she pauses again. She can point to the leaf but cannot yet explain its role in sustaining life. The diagram is accurate. The reasoning beneath it is still forming.

So, the teacher changes the conditions under which knowledge must live.

Instead of reviewing definitions, she places small pots of soil along the windowsill; some in steady light and others in shadow. Students plant fast-growing seeds. Some plants rotate daily. Before planting, students record predictions about what they believe will happen and why.

Days pass. The plant near the window bends toward the light. The shaded plant grows thin and pale. The rotated plant twists with each change in position. When a seedling is lifted during replanting, its roots reach downward regardless of how the pot was turned.

The teacher asks again: "Why do roots grow downward?" This time the answer is slower – and steadier. The student explains gravity and anchoring. She connects stems to light, light to energy, energy to survival. The labeled parts now belong to a living system. The worksheet was not wrong. It was incomplete.

Understanding deepens when students test ideas against reality – when naming becomes explaining, and explaining becomes reasoning. In classrooms where information is increasingly accessible and instantly generated, the discipline of watching something grow – observing, revising, reconciling prediction with evidence – keeps knowledge honest.

Correct answers identify. Investigation reveals.

Illustration in the Classroom: Middle School Level

A student has identified the central theme of the short story.

Textual evidence is cited precisely. Each comprehension question

is answered with efficiency. On paper, the expectation appears met.

Then a classmate raises a hand. "I don't think that's the theme at all."

The room shifts – not because someone is wrong, but because certainty meets resistance.

When asked to defend his interpretation, the student repeats phrases from the text. When pressed to explain how the character's decision shapes the outcome, his reasoning narrows. The answers remain technically correct. The connections between them are thin.

Rather than supplying clarification, the teacher rearranges the work of thinking.

Students move into triads. One defends an interpretation. One counters it using alternative evidence. One listens for what both readings overlook. They must identify not only what the theme is, but what evidence would destabilize it. If the setting were changed, would the theme endure? If a single event were removed, would the meaning shift?

Conversations slow. Interpretations bend under scrutiny and are reshaped. Revisions now happen mid-sentence. Students acknowledge competing ideas. Students integrate peer's counterexample.

Fluency begins to give way to explanation.

A summary can name a theme. Understanding requires holding multiple plausible meanings long enough for one to earn its claim through evidence. In that friction – between assertion and counterargument – thinking becomes visible.

Answers conclude. Reasoning continues.

Illustration in the Classroom: High School Level

The business plan is polished.

Market analysis is articulated with confidence. Revenue projections are formatted cleanly. Branding strategy, operational timeline, projected costs – all presented with precision.

Then the constraints arrive. A supplier withdraws unexpectedly. A competitor reduces pricing. A regulatory change alters operational costs.

The instructor asks for an assumption map. Every projection must be traced: Which numbers are supported by data? Which are inferred? Which rely on optimism? For each assumption, the student must identify what would cause it to fail.

When the conversation turns to projected revenue, the student references the table – but struggles to articulate why that growth rate was chosen. When pressed on customer behavior in the local market, responses grow tentative. The formatting remains strong. The judgment beneath it is still developing.

The exercise shifts from presentation to adaptation. Students revise projections under new constraints. Financial models include ranges rather than single outcomes. Contingency strategies are written into the plan. Risk is not hidden – it is analyzed.

The plan grows less pristine – and more credible.

A tool can generate a professional document.

Only disciplined reasoning can sustain it when reality shifts.

Illustrations in the Classroom: University Level

Students synthesize peer-reviewed research and propose implications for district policy. One paper stands out immediately.

The prose is elegant. Citations are embedded with sophistication. The argument unfolds with coherence and authority. It reads like publishable work.

During discussion, the teacher asks, "How would this equity framework alter decision-making when state accountability metrics prioritize standardized growth targets?" There is a pause.

The student begins by restating language from the paper. Equity should guide policy. Districts should respond more inclusively. The terminology is accurate. The application is uncertain.

Rather than pressing with more questions, the teacher restructures.

She introduces a policy-tension protocol. Each student must identify: the central claim, the constraint that challenges it, the trade-off that emerges, the unintended consequence that might follow, and a revised position in light of those pressures.

When confronted with trade-offs – between culturally sustaining practice and fixed funding formulas, between local autonomy and state compliance – the conversation thickens. Arguments falter, then reassemble. One student revises his position after recognizing that increased flexibility may reduce comparability. Another acknowledges that equity initiatives may require resource redistribution that invites political resistance.

By the end of discussion, the room sounds different. Language is less ornamental, more accountable. Students anticipate counterarguments before they are raised. They speak not from citation alone, but from judgment formed in dialogue.

Fluency is no longer mistaken for ownership.

A document can be generated. Judgment must be practiced.

Students can appear confident. They can produce correct answers. They can move quickly. Yet understanding often remains fragile until teachers pause, probe, and require explanation.

The illusion of knowing thrives on speed and polish. The art of teaching interrupts it.

And in an age where artificial intelligence can generate correctness effortlessly, that interruption is no longer optional – it is essential.

AI intensifies this tension. It delivers correct answers without struggle or reflection. While speed and correctness can impress evaluators, they do not guarantee understanding. AI produces the appearance of knowing without the labor of learning. It delivers fluent language, polished reasoning, and correct answers – instantly. The output looks like understanding, and to an untrained eye, it can be indistinguishable from it.

Correctness is not the same as significance. Producing the right answer does not guarantee an understanding of why it works – or why it matters.

Artificial intelligence can approximate correctness, but it cannot cultivate wisdom or true sense-making. As Davis and Marcus (2023) observe, "Current AI systems lack robust models of the world; they do not possess genuine understanding of the content they generate" (p. 7). Their fluency can be striking, their reasoning polished, yet what appears as thought is often simulation. Shanahan (2024) reminds us that large language models "simulate reasoning rather than instantiate it" (p. 15). The distinction is subtle but profound. AI is constrained by the quality and limits of the data that formed it – a digital echo of the long-standing principle of "garbage in, garbage out." If flawed information shapes the system, flawed outputs follow. The AI system cannot rise above what it has been fed. The quality of the input determines the integrity of the result.

When inputs are biased, shallow, or incomplete, outputs inevitably reflect those same distortions. Wisdom, by contrast, requires grounded judgment, lived experience, and moral discernment.

AI can produce answers. It cannot grow understanding.

Because AI depends on input quality rather than lived experience, it cannot replicate the human processes through which understanding is formed.

Where artificial systems rely on data patterns, human understanding grows through relationships. As Vygotsky (1978) explains, "Every function in the child's cultural development appears twice: first, on the social level, and later, on the individual level" (p. 57). Teaching lives in that movement from shared experience to internal meaning. Rogoff (2003) likewise reminds us that "Children's cognitive development is inseparable from the social and cultural activities in which they participate" (p. 11). Understanding is not uploaded; it is cultivated through participation, guidance, and presence.

Educators who guide students through confusion resist the illusion that speed equals intelligence. They create space for productive struggle, asking students to explain their reasoning, revise their thinking, and wrestle with uncertainty. As Dweck (2006) writes, "The passion for stretching yourself and sticking to it, even (or especially) when it's not going well, is the hallmark of the growth mindset" (p. 7). In this view, difficulty is not evidence of limitation but an invitation to grow, for "effort is what ignites ability and turns it into accomplishment" (Dweck, 2006, p. 41). These moments of sustained effort matter. Bandura (1997) reminds us that "Mastery experiences are the most effective way of creating a strong sense of efficacy" (p. 80). In the space between confusion and clarity, problem-solving, perseverance, and self-belief are formed. The moment we honor struggle, we collide with competing priorities.

This places our education system, artificial intelligence, and the art of teaching at a critical point of tension.

AI exposes a cultural misunderstanding long tolerated in education: that learning is about producing correct output. Yet, as Perkins (1992) reminds us, "Education aims at understanding, not just knowledge" (p. 3).

Learning is fundamentally about transformation. It changes how a person sees the world and themselves within it. It reshapes judgment, empathy, and agency. These shifts are invisible on worksheets and untraceable in algorithms, yet they are essential to understanding. As Hattie (2009) writes, "Teaching is not about the transmission of knowledge, but about creating conditions for learning" (p. 243). This is why students can generate essays without defending arguments, solve problems without grasping implications, or repeat language without internalizing it. Correctness can be displayed; understanding must be enacted.

There is an inherent time constraint to school system's funding model that does not account for the learning process. To that end, the art of teaching is boxed into the school calendar, curriculum scope and sequence, employment contracts, and bargaining agreements. Yet while systems impose rigid timelines and structures, the process of true understanding does not conform to schedules or mandates; it unfolds at its own pace within the learner.

So, how do we foster the art of teaching in a system where teacher evaluations are measured, at least in part, by student performance metrics or correct answers? How do we grow student understanding in a system where butts in seats equal teaching jobs?

The answer begins with remembering how learning actually works. We must return to the nature of understanding itself.

Human understanding develops over time. It forms through struggle, contradiction, revision, and reflection. It is shaped by context and experience. When a learner truly understands something, they can adapt it, critique it, and apply it in unfamiliar situations. They can explain not just what works, but when and why it might not.

AI cannot do this. It does not understand truth – it approximates it.

This is why access to information has never guaranteed wisdom. Libraries have always existed, whether in spoken traditions or buildings full of books. So has misinformation, bias, and shallow thinking. The challenge was never scarcity of knowledge. The challenge was sense-making. Teaching lives in that space.

It would be a simple job to give students the right answers. That is what AI does.

When teachers ask students to explain their thinking, revise their ideas, or wrestle with uncertainty, they are teaching something far more durable than content. They teach both problem-solving and sense-making alongside grit, perseverance, and stick-to-it-ness.

The danger is not that AI will make students less intelligent. The danger is that it reinforces a shallow definition of intelligence – one schools should have abandoned long ago. While AI can replicate the production of correct answers, it cannot replace the relational art of teaching: performance, guidance through productive struggle, encouragement, and human connection. As Pianta and colleagues (2012) explain, "Teacher–student interactions are the primary mechanism through which teachers influence student engagement and achievement" (p. 365). It is the quality of those interactions, not the quantity of instruction, that shapes growth.

Understanding flourishes in the soil of trust. Wentzel (2010) found that "Perceived teacher caring and support are consistently linked to students' motivation and academic engagement" (p. 76). Intelligence grows where relationships make risk possible.

Teachers remain the guardians against the illusion that learning is synonymous with correctness. They notice when confidence masks confusion, listen for reasoning beneath correctness, and cultivate understanding through scaffolding and interaction. As Hattie (2009) reminds us, "It is teachers who have the greatest effect on student learning" (p.238).

In an age where answers are abundant, understanding becomes a rare skill, and teaching becomes more precise. AI may expose gaps in the art of teaching, but it cannot replace the qualities teachers bring to learning: mentoring curiosity, modeling critical thinking, fostering ethical reasoning, and supporting students as they construct meaning in their own lives.

The task of educators is not to simply provide information but to guide students in discerning, questioning, testing, and connecting knowledge to lived experience.

AI can provide answers; teachers cultivate understanding.

One may impress the eye, but the other endures for a lifetime.

Invitations for Preservice Teachers

Preservice Teachers

Correct answers can be convincing.

They are not proof of understanding.

One of the most important professional habits you will develop is learning to see beyond performance. Students may produce polished responses, fluent explanations, and accurate solutions. In an AI-driven world, they may even generate sophisticated work with minimal effort. Your task is not to be impressed by fluency; it is to probe for reasoning.

Ask deeper questions: Can students explain their thinking? Can they transfer knowledge to unfamiliar contexts? Can they defend their ideas when challenged – or revise them when confronted with new information? These are indicators of understanding.

In an era where AI can produce answers instantly, your greatest contribution is not speed or efficiency. It is discernment. You help students slow down long enough to wrestle with ideas, confront misconceptions, and construct meaning for themselves. Understanding develops more slowly than output, and your patience – your willingness to pause, question, and guide – is part of the learning itself.

You are invited to consider: When a student's work appears polished, do you feel relief – or curiosity? Do your questions press for transfer and structure, or settle for accuracy? Where might you unintentionally equate confidence with comprehension?

Train yourself to question fluency. When students appear confident, probe for structure. Ask what assumptions underlie their answer and whether their reasoning transfers to a new context. Your role here is epistemic – not motivational. You can resist the illusion that correctness equals mastery. And that resistance will define the kind of teacher you become.

Invitations for Administrators

Challenge long-standing assumptions.

Many exist embedded in educational systems.

There is an assumption that visible performance is synonymous with learning. While test scores, completion rates, and polished student work are easy to measure, understanding is not.

Artificial intelligence intensifies this tension. When tools can generate fluent responses instantly, surface-level performance becomes even less reliable as an indicator of comprehension. AI exposes how frequently schools reward correctness over conceptual depth – not because educators are careless, but because systems are designed to privilege what can be counted. Consider what your systems truly reward.

You are invited to consider: When you review data dashboards and performance metrics, what remains invisible? Do your structures honor revision and reasoning – or merely accuracy and speed? Where might your accountability measures unintentionally privilege polish over durability?

Leadership in this moment requires clarity and courage. Administrators must examine whether assessment policies, evaluation frameworks, and accountability measures prioritize explanation, revision, and transfer – or simply correct answers. Protecting time for productive struggle, dialogue, and reflective thinking is not inefficiency; it is fidelity to how learning actually occurs.

Examine whether your systems confuse polish with understanding. What metrics in your school reward completion without conceptual durability? Redesign evaluation structures to privilege explanation, not just performance. Send a powerful message: understanding matters more than appearance. In an AI-saturated landscape, that distinction is no longer philosophical – it is foundational.

Invitations for Homeschool Parents

It is tempting to equate quick completion with mastery.

When your student answers quickly, resist the urge to move on. Ask them to teach it back, to solve it differently, or to explain what would make the answer fail. Understanding reveals itself when it is stretched.

Ask your student to describe their reasoning. Invite them to teach the concept back to you. Present a slightly altered situation and see whether they can adapt to what they know. Provide them with solutions to solve, errors to find, and encourage them to verify. These small shifts transform learning from recall to reflection. Consider how often you linger in the space between answer and understanding.

You are invited to consider: When your student responds quickly, do you interpret speed as mastery? Where might you slow the moment just enough for deeper thinking to surface? Do your routines reward completion – or comprehension?

Artificial intelligence may provide answers quickly. It may even generate convincing explanations. But understanding is formed in conversation – in the back-and-forth exchange where questions are asked, confusion is surfaced, and thinking is revised. When you linger in those moments, you help your student move beyond the illusion of knowing and into genuine comprehension.

In a world where answers are abundant, your role is not to supply more of them. It is to cultivate discernment, curiosity, and the ability to explain why something matters. That is the kind of learning that endures.

Reflection Questions

Chapter 4: The Illusion of Knowing

1. Where in your classroom might correctness be mistaken for comprehension?

2. What do your assessment structures reward more: explanation or efficiency?

3. When was the last time you paused a lesson to probe beneath a correct answer?

4. If a student used AI to complete a task, would shallow understanding be exposed – or protected?

5. Where in your classroom must students defend, revise, or transfer their thinking under pressure?

Chapter 5: The Emotional Architecture of Learning

Learning does not begin with information. It begins with safety.

Before a student engages with content – before curiosity, effort, or persistence can take hold – the brain makes a faster, quieter assessment: Am I safe here? Do I belong? Will I be seen or exposed? These questions are not distractions from learning. They are prerequisites.

Decades of research in psychology, neuroscience, and education show that social and emotional contexts shape attention, memory, and problem-solving. As Immordino-Yang and Damasio (2007) explain, "The aspects of cognition that are recruited most heavily in school… are profoundly affected by emotion" (p. 3). Emotion is not an interruption to learning; it is woven into its architecture. Similarly, Blum (2005) notes "Students who feel connected to their school are more likely to engage in academic tasks and achieve at higher levels" (p. 231). The science has been clear; our structures, less so.

Traditional schooling has often reacted to disruptive behavior, treated compliance as evidence of learning, and relegated emotion to the margins – even though it sits at the center of how learning actually occurs. "Emotion is integral to the processes of reasoning and decision making" (Immordino-Yang & Damasio, 2007, p.10).

For generations, education treated emotion as an accessory to cognition – something to manage after instruction rather than something that shapes it from the start. Classrooms were designed around efficiency, standardization, and control, as if learning could be separated from feeling. But the human mind has never separated thought from feeling; cognition has never functioned independently of emotion. Humans are not machines.

Neuroscience now confirms what teachers have always known intuitively: emotion and cognition are inseparable. "We only think deeply about things we care about" (Immordino-Yang, 2016, p.18). Stress narrows thinking, while belonging broadens it; curiosity opens neural pathways that fear closes. "Anxiety can impair working memory and reduce cognitive flexibility" (Pekrun, 2006, p.324). Motivation, attention, and reasoning are not neutral processes – they are filtered through emotional states. Students who feel anxious, marginalized, or invisible process information differently than those who feel safe, respected, and valued.

There was a morning when the emotional architecture of a room revealed itself to me in ways I could not ignore. The students entered more quietly than usual. Their movements were tight, shoulders raised, eyes scanning. I had been informed that there had been a physical altercation in one of the living units the night before – an injury severe enough to require medical attention. I registered the information as context, not as climate.

I began the lesson as planned.

My voice carried more urgency than usual. I pressed for responses. I moved quickly through the objectives, determined not to let disruption steal instructional time. But the room would not follow. Eyes drifted. Pens remained still. When I called on students, answers were clipped or absent. What I interpreted as defiance was, in truth, distance.

It took me too long to realize that their nervous systems were elsewhere. Mine was, too.

My own heightened state – subtle but present – had tightened the air further. I mistook emotional aftershock for disengagement. I mistook silence for resistance. In my effort to preserve rigor, I had failed to attend to regulation.

The lesson did not fail because the content was unclear. It failed because the emotional ground beneath it was unstable.

That day altered how I enter rooms shaped by unseen events. Learning does not resume simply because the bell rings. Safety cannot be assumed because chairs are filled. Before cognition can expand, the nervous system must settle. And when mine is unsettled, theirs rarely steadies.

Artificial intelligence does not experience any of this.

AI can generate explanations without caring whether anyone is confused. It can respond without noticing hesitation, tone, or shame. It does not register when a student disengages because the risk of being wrong feels too high.

Teachers do.

They notice micro-expressions, shifts in posture, and subtle cues of uncertainty. They detect when confusion is masked or when peer dynamics silence voices. This contrast reveals a fundamental truth: while AI can deliver content flawlessly, it cannot replicate the human sensitivity and responsiveness that make learning both meaningful and safe.

AI can generate explanations with precision and speed, but it does not register hesitation in a student's voice, nor does it recalibrate when confusion flickers across a face. Even my own tendency to say or type "please" when making a request to an AI system exposes the illusion – we may feel courteous, but the system remains unchanged, unaffected by gratitude or respect. Politeness toward a machine alters little because the machine does not participate in relationship.

By contrast, human interaction is shaped continuously by nuance: a softened voice lowers anxiety, a pause invites courage, a raised eyebrow can either shame or signal curiosity. These subtle shifts accumulate into relational safety. And it is only within that safety – constructed moment by moment through attention, tone, and attunement – that learning moves beyond information and becomes transformative.

The Invisible Architecture of Learning

Every classroom operates on an emotional architecture – an invisible structure built from relationships, expectations, trust, and responsiveness. Some environments invite risk-taking, while others reward silence and compliance. Some classrooms communicate, "You can try here," while others whisper, "Don't mess up."

Research in social-emotional learning demonstrates that these signals – often transmitted in seconds – shape engagement and achievement more powerfully than lesson plans alone. As Zins and colleagues (2004) note, "There is growing evidence that social and emotional competencies are associated with academic achievement" (p. 6). Likewise, "Classrooms characterized by emotionally supportive interactions show higher levels of student engagement" (Pianta et al., 2012, p. 368). When students feel safe, they ask questions. When they feel seen, they persist. When they feel respected, they stretch beyond what feels comfortable.

These effects accumulate through thousands of micro-decisions teachers make daily: how they respond to mistakes, whose voices they elevate, and how they navigate uncertainty. They are visible in the pause a teacher gives before calling on a hesitant student, the choice to say "Tell me more" instead of "That's not quite right," the decision to validate effort before correcting reasoning, and the subtle redirection that preserves dignity rather than exposing error. Over time, these subtle acts shape who feels invited to learn.

These effects also appear in how teachers frame group work – intentionally pairing students for growth rather than convenience – how they pronounce a student's name carefully and correctly, how they redistribute airtime when certain voices dominate, and how they acknowledge cultural references or lived experiences offered in discussion. In these gestures, the invisible climate of learning takes shape.

These effects are present in the choice to kneel beside a desk instead of correcting from across the room, to allow wait time instead of filling silence, to check in privately rather than publicly, and to treat a behavioral misstep as information rather than defiance.

These moments, small as they seem, become the framework of belonging.

None of these decisions require a new curriculum, yet together they construct the emotional architecture of the classroom. Over time, these small, relational calibrations communicate a powerful message: you are safe here, your thinking matters, and your growth is worth patience.

The art of teaching lives not in grand gestures, but in accumulated attentiveness.

Illustration in the Classroom: Elementary Level

The materials are simple: baking soda, vinegar,

clear containers, measuring cups.

Before combining them, the teacher asks students to write a prediction and explain their reasoning. A student stares at his paper. He has watched experiments before. He has seen classmates answer quickly and confidently. His hesitation is not about chemical reactions. It is about exposure. What if he is wrong in front of everyone? What if his prediction sounds naïve? What if the class laughs? His pencil hovers.

The teacher does not rush the silence. Instead, she reframes the moment. "Predictions are not promises," she says. "They are starting points." She writes her own on the board: I think it might bubble slowly – but I'm not certain. Then she circles the words...not certain. She does not erase uncertainty. She names it.

Students begin writing again. Some predictions are cautious. Some are bold. When the mixture erupts more dramatically than expected, the room fills with startled laughter. The teacher does not correct mismatched predictions. She asks, "Whose thinking changed? What surprised you?"

In this classroom, being wrong is not public failure. It is participation in discovery. Over time, predictions grow less guarded. Students explain their reasoning more fully. Unexpected results are greeted with curiosity rather than embarrassment.

The experiment is simple. The structure surrounding it is not.

Scientific thinking requires risk – the willingness to offer a hypothesis before evidence is secure. That willingness does not emerge from content alone. It grows from tone, modeling, pacing, and the subtle cues that tell a child whether uncertainty is safe.

In this room, courage is rehearsed alongside chemistry.

Illustration in the Classroom: Middle School Level

Students are analyzing primary-source documents from a

contentious historical moment. The text is dense.

The implications are layered. Interpretations carry social weight.

One student reads carefully but rarely speaks. When discussion opens, his eyes move first – not to the document, but to his peers. He is measuring the room. Intellectual risk is rarely solitary. It is social.

Offering an interpretation means risking correction. Challenging a peer's claim can feel like confrontation. Silence can feel safer than misreading history in front of classmates.

The teacher does not address this tension through reassurance alone. She addresses it structurally. Discussion protocols shift. Students begin with written claims before speaking. Small groups generate two competing interpretations and must articulate the strongest evidence for each. Roles rotate: one student advances a claim, another strengthens it, a third identifies possible counterevidence. Disagreement is rehearsed before it is performed publicly.

The teacher models how to question without diminishing: "What evidence leads you there?" rather than "That's incorrect." When a student revises a claim mid-discussion, the adjustment is named as strength. "Notice what just happened," she says. "That's historical thinking."

Over time, participation shifts. Not louder – but steadier. Students challenge ideas with textual evidence. They refine interpretations when new information surfaces. They begin to treat ambiguity not as threat, but as terrain to navigate. Emotional architecture at this age is not about eliminating tension. It is about structuring, so it becomes productive.

Analysis deepens when students trust that revision does not diminish their standing in the room.

Illustration in the Classroom: High School Level

Students publicly defend positions on a current policy issue.

They balance constitutional rights, public safety, competing interpretations of precedent. Arguments are constructed in advance. Rebuttals are anticipated. The reasoning is visible. So is the vulnerability.

When a student begins presenting, others prepare counterarguments. A misstatement can unravel under scrutiny. An underdeveloped claim can invite rapid correction. The line between intellectual challenge and personal exposure is thin.

The teacher understands that civic reasoning requires both rigor and relational safety. Before debate begins, norms are not simply announced – they are practiced. Students rehearse how to challenge ideas without assigning motive. They practice paraphrasing before rebutting. They learn to ask clarifying questions before advancing critique.

During debate, the teacher listens as carefully to tone as to logic. When tension sharpens into edge, she pauses – not to silence disagreement, but to recalibrate. "Let's return to the argument," she says. "What principle is being contested here?"

When a student adjusts a position after hearing a counterpoint, that movement is named explicitly: "That revision shows you are thinking."

Gradually, the quality of engagement shifts. Students prepare not only to defend, but to interrogate their own reasoning. They anticipate counterarguments. They acknowledge trade-offs. They distinguish between personal conviction and constitutional interpretation.

In this classroom, safety is not the absence of disagreement. It is the assurance that identity is not at stake when ideas are revised.

Civic discourse matures when students learn that changing one's mind in light of evidence is not weakness – but discipline.

Emotional architecture determines whether learning remains superficial compliance or becomes meaningful engagement. The ability to struggle safely – to experiment, reflect, and revise – is what allows understanding to deepen and endure.

These dynamics surface in ordinary classroom moments where safety and trust shape how students engage. From hesitating to read aloud, to fearing misinterpretation in discussions, to navigating labs, engagement is mediated by safety, trust, and relational support. In each case, teachers scaffold learning by modeling thinking, normalizing mistakes, validating effort, and providing opportunities for reflection, creating environments where curiosity, risk-taking, and persistence is possible.

Relational interventions transform fear into confidence, isolation into participation, and tentative reasoning into deep understanding. "Emotionally supportive classrooms promote students' willingness to take academic risks" (Pianta et al., 2012, p. 372). When students trust their own capacity, they remain in the struggle long enough for understanding to take root. The teacher's role, then, is to "diagnose and respond to students' misconceptions" (Hattie, 2009, p. 238). In these ways, learning is inseparable from emotional context.

Teaching becomes most visible in moments of struggle.

It is easy to appear competent when answers come quickly. It is harder – and far more revealing – when thinking falters. Failure is not merely academic; it is psychological. It poses a dangerous and deeply personal question: What does this say about me? In that moment, students are not only evaluating their work; they are evaluating their worth.

Teachers stand in that space. They help students reinterpret failure – not as identity, but as information. As Dweck (2006) writes, "In the fixed mindset, everything is about the outcome. If you fail... it's all been wasted" (p. 6). Yet when students come to believe that ability can grow, "they are more resilient in the face of setbacks" (Yeager & Dweck, 2012, p. 303). Through tone, posture, patience, and consistency, teachers communicate – often without explicit language – that growth is possible and that struggle does not diminish belonging. In doing so, they protect students from internalizing shame and instead cultivate resilience.

Artificial intelligence cannot replicate this work. It can generate answers, refine explanations, and correct errors, but it does not perceive fear. It does not recognize the silence that follows embarrassment. It cannot sense the tightening shoulders, the averted gaze, or the fragile confidence behind a hesitant voice. It cannot repair relational ruptures or restore trust after a moment of vulnerability. AI cannot build emotional architecture because it does not participate in relationship.

Teaching, at its best, is emotional labor disguised as instruction. It is the quiet work of stabilizing identity while advancing intellect. It is the steady reassurance that confusion is survivable and that mistakes are steppingstones rather than verdicts.

As AI becomes more integrated into classrooms, the emotional role of the teacher does not diminish – it intensifies. Students must now navigate not only content, but identity: who they are as thinkers in a world where machines produce flawless answers instantly.

Without intentional emotional scaffolding, students may outsource thinking – not from laziness, but from fear. Fear of being wrong. Fear of being slower than the algorithm. Fear of discovering that their effort cannot compete with perfection. Teachers counter that fear by redefining the standard. Humanity – not perfection – is the measure of growth.

AI can correct errors. It cannot interpret them with compassion. It cannot respond to them with authentic encouragement. It cannot look a student in the eye and help them believe they are still capable.

Emotion is not an accessory to learning; it is its foundation. In classrooms that prioritize emotional architecture, students develop persistence, judgment, empathy, and self-efficacy alongside academic skill. They learn how to navigate complexity without collapsing into self-doubt. They discover that courage is not the absence of struggle, but the willingness to continue through it.

AI may provide answers. Teachers cultivate courage. And in an age increasingly obsessed with intelligence, teachers remain the architects of something even more powerful: resilience.

Invitations for Preservice Teachers

Before students can learn, they must feel safe enough to try.

Emotion is not separate from cognition – it shapes it. Attention, memory, motivation, and reasoning are filtered through a student's sense of belonging and security. If the brain is scanning for threat, it cannot fully engage with content.

As a preservice teacher, you are not simply planning lessons; you are constructing emotional architecture. Your tone when correcting a mistake, your posture when a student hesitates, your consistency in expectations, and your willingness to normalize confusion all send signals. Those signals accumulate. They determine whether students interpret struggle as growth or as exposure.

You are invited to consider what your presence communicates before your lesson ever begins. In an AI-saturated world, where students can access instant answers, your role becomes even more defined. You are not competing with machines for speed or efficiency. You are cultivating trust. You are modeling how to think through uncertainty without collapsing into self-doubt. You are helping students build the courage to persist when learning feels messy.

You are invited to consider: When students enter your room, what do they feel first? Do your responses lower threat or heighten it? When mistakes surface, does your posture signal impatience – or possibility? What emotional patterns are you rehearsing daily through tone, timing, and attention?

Your responsiveness is not a personality trait – it is pedagogy. The safety you create becomes the foundation upon which understanding is built. Emotional architecture is not an optional enhancement to instruction; it is the structure that makes instruction possible.

Invitations for Administrators

Policies communicate values.

Learning environments are built emotionally long before they are measured academically. Examine the invisible architecture of schools: norms, pacing expectations, evaluation systems, and accountability pressures that shape psychological safety. Tight pacing guides can signal that speed matters more than depth. Public performance metrics may elevate correctness over growth. Evaluation frameworks may reward polish while overlooking the relational work required to build trust. These choices influence whether classrooms feel like spaces of risk-taking or surveillance. Compliance remains necessary, but school improvement is strengthened when leaders intentionally attend to emotional architecture.

You are invited to consider: What do your systems communicate about struggle? When teachers take instructional risks, do they feel protected or exposed? When students falter, do your structures interpret that as data for growth—or as deficiency to correct? Where might your policies unintentionally prioritize performance over psychological safety?

When administrators intentionally prioritize belonging, relational trust, and emotional safety, they create the conditions for intellectual risk. These commitments do not compete with academic outcomes; they strengthen them. In psychologically safe environments, teachers feel empowered to experiment, revise instruction, and engage in reflective practice. Students learn that questions are welcome, confusion is part of learning, and persistence matters more than perfection.

Artificial intelligence heightens this leadership responsibility. As AI makes polished answers easier to produce, schools must intentionally protect depth. Leaders set this tone. By safeguarding time for explanation, revision, and productive struggle, they preserve both instructional quality and the emotional architecture that sustains it. Psychological safety is not peripheral to academic excellence—it is foundational.

Invitations for Homeschool Parents

One of homeschooling's greatest strengths is emotional safety.

When students feel secure, seen, and supported, learning unfolds more naturally. Emotional architecture at home is not abstract; it is lived daily in tone, routine, and relationship.

When a student struggles with math, reading, or writing, your response communicates more than feedback – it communicates identity. Do mistakes signal failure, or do they signal growth? Is confusion met with frustration, or with curiosity? These daily interactions shape how your student understands both learning and themselves.

You are invited to consider: In moments of struggle, what does your tone teach? Does your student experience correction as criticism – or as care? When uncertainty lingers, do you rush to resolve it, or remain present within it?

In a world where AI can provide answers instantly, the deeper work remains relational.

Understanding develops through conversation – through asking why, revisiting thinking, and reflecting together. When you invite your student to explain their reasoning, revise their ideas, and tolerate uncertainty, you reinforce that mistakes are informative, not threatening.

Emotional architecture is not invisible – it is embodied. It shapes whether your student approach challenges with avoidance or resilience. Intellectual growth is rooted in emotional safety. When trust is present, curiosity expands. When belonging is secure, risk-taking follows.

Tools may assist learning. Relationship sustains it.

Reflection Questions

Chapter 5: The Emotional Architecture of Learning

1. What emotional states does your classroom design consistently produce – and how do you know?

2. Where might your grading, pacing or participation structures be activating threat rather than growth?

3. When have you witnessed a student think more clearly because they felt seen or safe?

4. How do you distinguish between productive discomfort and emotional overload in your classroom?

5. Do students in your classroom learn how to regulate emotion during challenge – or simply avoid challenge altogether?

PART II — MEETING AI WITHOUT FEAR

Chapter 6: What AI Actually Is (and Isn't)

Learning is human before it is intellectual. It is built on safety, relationship, and the emotional architecture that allows risk-taking and growth. Artificial intelligence cannot construct that architecture because it does not participate in relationship. Yet much of the fear surrounding AI in education stems from attributing to it qualities it does not possess – understanding, intention, even consciousness. To move forward thoughtfully rather than reactively, we must separate what AI cannot do from what it actually does.

Fear thrives in vagueness. Clarity restores proportion. Before deciding how AI should fit within education, we must first understand what it is – and what it is not.

AI does not think. It does not understand. It does not intend. It predicts.

At its core, AI identifies patterns in vast amounts of data and generates responses based on probability. When it produces language that feels thoughtful or insightful, it is mimicking human patterns – not because it understands meaning, but because it has detected statistical regularities in how language is used.

This distinction matters.

Humans learn by connecting experience, emotion, and reflection. AI "learns" by correlating inputs and outputs. "Computers are good at correlations. They are not good at understanding" (Broussard, 2018, p. 17). They do not possess memory in the human sense, nor do they have judgment, conscience, or curiosity. "Current AI systems do not have consciousness or self-awareness" (Russell and Norvig, 2021, p. 1025). And yet, because AI's outputs sound fluent, we project intelligence onto it – mistaking pattern generation for thought.

Students may treat AI-generated explanations as "smart" or inherently authoritative in much the same way we instinctively anthropomorphize the technologies that surround us. We casually say our phone "doesn't want to work," our GPS "chooses the wrong path," or our printer "refuses to cooperate," as though intention resides inside circuitry. We scold a smart speaker for mishearing us, assigning motive where there is only code.

When a system behaves in ways that appear purposeful, we naturally infer the presence of a mind behind it. That impulse is deeply human – and largely harmless in everyday life – but in educational contexts it can quietly distort judgment.

If we equate fluency with thoughtfulness or responsiveness with comprehension, we risk mistaking statistical pattern recognition for genuine understanding. We often predict and explain behavior "whether or not the system really has beliefs" (Dennett, 1987, p. 49). Decades after this quote was written – in a world transformed by computational power and intelligent-sounding machines – the insight remains strikingly consistent. Or more simply put, "Today's AI systems do not understand the world; they manipulate symbols" (Hawkins, 2023, p. 112).

Recognizing the difference between simulated intelligence and lived cognition is therefore not optional; it is a critical literacy skill for both students and teachers navigating an age of intelligent-sounding machines.

Illustration in the Classroom: Elementary Level

The student reads the paragraph aloud with steady confidence.

It is well organized. The sentences transition smoothly. Key details from the historical event are included. He used an AI program to generate the summary.

The paragraph answers the question it was asked to answer. Then the teacher changes the question. "Why did this event matter?"

He looks back at the page. He repeats a sentence. Then another. The wording is precise. The explanation remains distant. When asked how the event changed people's lives – or what might have happened if it had unfolded differently – his responses narrow.

The summary is accurate. The meaning has not yet settled.

The teacher does not remove the tool from the equation. Instead, she reframes its role. "Let's treat this as a first draft," she says. "Now let's see what it leaves out."

Together they return to the original text. They circle moments the summary compresses. They ask: Who benefited? Who was harmed? What consequences unfolded over time? The student adds marginal notes. He underlines cause-and-effect relationships. He writes a second paragraph – not replacing the first, but extending it.

The AI-generated summary becomes scaffolding, not substitute.

By the end of the lesson, his explanation sounds different – not because it is more complex, but because it carries ownership. He connects actions to consequences. He references specific people affected. He adds a question the summary never raised.

The tool produced language. The thinking required interrogation.

Illustration in the Classroom: Middle School Level

The explanation of Newton's laws is thorough.

Definitions are precise. Diagrams are labeled. Calculations align.

The student generated the explanation using AI before class.

When asked to predict how friction would affect a skateboarder on different surfaces, she begins by reciting: "Force equals mass times acceleration." The equation is correct. The application hesitates.

So, the teacher changes the medium.

Instead of revisiting definitions, students move into motion trials. They measure distance and time across carpet, tile, and concrete. They record data before calculating acceleration. Predictions are written first – without devices open.

After the trials, the AI explanation is projected onto the board. The class annotates it collectively.

Where does the explanation assume ideal conditions? Where does it oversimplify friction? Which variables did our experiment reveal that the explanation ignores?

The text becomes something to critique.

The student who had relied on the generated explanation begins circling phrases that now feel incomplete. She adds handwritten notes in the margins: surface texture matters, energy lost to heat, real-world friction varies.

The principle moves from memorized formula to adaptable reasoning.

Artificial intelligence produced a clean explanation. The experiment revealed its limits. Neither is sufficient alone.

Illustration in the Classroom: High School Level

A complex poem has been interpreted with stylistic sophistication.

The essay is elegant. The thesis is coherent. Metaphors are identified. Historical context is referenced. Citations are formatted correctly. The student used AI to draft the initial version.

Rather than questioning authorship, the teacher questions certainty. She projects a single line from the poem and asks the class to annotate it without prepared notes. Interpretations emerge slowly. One student sees isolation. Another sees defiance. A third points to irony.

The AI-generated essay had treated the metaphor as unambiguous. Now ambiguity fills the room. The student who submitted the polished essay begins defending the original claim, then pauses. A peer introduces an alternative reading supported by textual evidence. He rereads the line. He notices a phrase he had overlooked. His interpretation begins to stretch.

The teacher asks a different question: "Which parts of the essay came easily? Which required the most revision?" The student reflects. "The structure was clear," he says. "But I didn't think deeply about that stanza." The discussion shifts from product to process.

Over time, his written work changes – not less fluent, but more annotated. Claims are accompanied by counterarguments. Interpretations acknowledge ambiguity rather than smoothing it away.

Artificial intelligence can assemble analysis from patterns across texts. It can predict plausible interpretations. It cannot experience the friction of competing readings in real time.

Discernment develops when students must defend, revise, and sometimes relinquish certainty.

The difference is not between human and machine. It is between performance and ownership

AI can produce fluent language, correct calculations, and coherent essays, but it cannot generate insight, judgment, or ethical reasoning. Meaningful learning depends on relational scaffolding, reflection, and intellectual struggle. Understanding emerges when students learn to move between representation and reality – between equation and observation – without confusing fluency for transfer.

AI can handle the mechanics of response generation. It cannot replace the human work of guiding students toward understanding, meaning, and growth. Prediction is not comprehension.

Schools do not exist to produce perfect outputs; they exist to cultivate thinkers, collaborators, and citizens.

AI can assist learning – it can generate examples, provide practice problems, and expand access – but it cannot determine what matters, contextualize failure, or guide identity formation. "Learning is shaped by the dynamic interactions among cognitive, social, and emotional processes" (Darling-Hammond et al., 2020, p. 98). Teachers navigate those interactions daily, often invisibly, shaping moral reasoning, ethical reflection, and metacognitive growth. They help students interpret setbacks, exercise agency, and understand who they are becoming in the process of learning – work that extends far beyond the production of correct answers.

Understanding what AI is allows educators to stop competing with it and start using it wisely. AI clarifies, rather than diminishes, the need for human judgment. When teachers embrace this perspective, they can focus on cultivating understanding rather than policing correctness. The danger is not that students will use AI, but that we will misunderstand learning so deeply that we forget why teaching exists at all.

Artificial intelligence amplifies, rather than diminishes, the need for human judgment.

Invitations for Preservice Teachers

Preservice Teachers

Your power as a teacher lives in the space where meaning begins.

Understanding what artificial intelligence actually is – and what it is not – liberates you from the quiet pressure to compete with it. When AI's output sounds confident, it can tempt learners – and educators – to mistake fluency for understanding. But polished responses are not proof of comprehension. Prediction is not judgment. Output is not insight.

You are invited to consider: Where might you be tempted to equate fluency with learning? When a response sounds sophisticated, do you pause long enough to ask whether it reflects genuine understanding? How will you help students recognize the difference?

Your work is not to outperform AI in speed or information retrieval. It is to help students slow down long enough to think. Ask them to explain their reasoning, to revise their claims, to wrestle with ambiguity. Guide them in distinguishing between a sentence that sounds intelligent and an idea that has been examined, tested, and internalized.

Teaching means cultivating intellectual discernment. Students must learn when to trust a tool, when to question it, and when to rely on their own reasoning. They must learn that understanding requires reflection and that ethical judgment requires context. Those capacities are not automated; they are formed relationally.

Learn to articulate clearly what AI does: it predicts patterns; it does not reason, judge, or care. Your instruction must begin with conceptual clarity. Students cannot use tools wisely if they misunderstand tools.

As you enter the profession, remember this: your humanity – your curiosity, responsiveness, moral awareness, and willingness to stay with students through confusion – is not something AI possesses.

Invitations for Administrators

Clarity creates the foundation for steady leadership.

Artificial intelligence can generate content, summarize data, and automate processes. It cannot evaluate understanding in context, interpret nuance, or make ethical decisions about what matters most in learning. When fluency is mistaken for intelligence, schools risk reorganizing themselves around output rather than understanding.

You are invited to consider: How is intelligence defined within your system? Do your assessments privilege speed and polish, or reasoning and transfer? What signals do your policies send about what truly counts?

If AI can produce correct answers easily, then systems must become more intentional about measuring explanation, judgment, and ethical reflection. The presence of AI does not diminish the need for professional judgment; it intensifies it. Leadership decisions about assessment, evaluation, pacing, and instructional expectations determine whether schools chase visible performance or cultivate durable understanding.

Avoid policies driven by anthropomorphism. If AI is treated as intelligence rather than computation, decisions will distort both risk and opportunity. Lead from definitional precision.

Thoughtful leadership in an AI era does not mean tightening surveillance or banning tools out of fear. Nor does it mean uncritical adoption of every emerging platform. It means protecting the integrity of learning. It means designing policies that reward depth rather than polish, and creating cultures where teachers are trusted to exercise judgment.

When administrators anchor schools in a clear understanding of learning – rather than in anxiety about technology – they steady the system.

AI may accelerate change, but principled leadership ensures that innovation serves human growth rather than redefining it.

Invitations for Homeschool Parents

When AI enters the home, it can feel impressive and unsettling.

It produces quick answers, coherent explanations, and seemingly authoritative responses.

AI may sound knowledgeable, but it does not understand meaning, context, or consequence. Help your student recognize that fluency is not wisdom. When a machine produces language that appears thoughtful, gently name what is missing – lived experience, accountability, moral responsibility.

You are invited to consider: What habits are forming as your student interacts with these tools? Do they accept responses at face value, or learn to question them? Are you modeling discernment in your own use?

You are uniquely positioned to cultivate discernment rather than dependence. You can pause after an AI-generated explanation and ask, "Does this make sense to you?" You can invite your student to restate ideas in their own words, apply them in a new context, or question assumptions embedded in the response. In doing so, you help your student move beyond passive consumption and toward active reasoning.

The goal is not to eliminate tools, but to frame them wisely. AI can provide information; it cannot model judgment. It can supply answers; it cannot teach when an answer is incomplete, biased, or ethically complicated. Those lessons emerge in conversation – in the back-and-forth exchange that clarifies thinking and strengthens agency.

In a world filled with intelligent-sounding systems, your presence teaches something enduring: how to think intentionally, how to question thoughtfully, and how to engage tools without surrendering autonomy. AI may assist learning, but it is relationships that anchor understanding.

Reflection Questions

Chapter 6: What AI Actually Is (and Isn't)

1. What assumptions about intelligence are you projecting onto AI that may not actually be true?

2. Where might you be equating fluent output with genuine comprehension – in machines or in students?

3. If AI can produce explanations instantly, what aspects of your professional judgment become more – not less – important?

4. Are your decisions about AI driven more by fear of replacement or clarity of purpose?

5. What you chose to use – or restrict – AI in your classroom, what kind of thinking are you protecting?

Chapter 7: The Panic Cycle in Education

Understanding that AI predicts rather than thinks should logically calm us. Instead, in education, clarity is frequently followed by urgency. When new tools disrupt familiar routines, fear often moves faster than design.

Every generation of educators has confronted an innovation believed to threaten learning itself. Socrates warned that writing would create the illusion of wisdom without true understanding – that that reliance on written text would weaken the mind's capacity to retain knowledge. Similarly, the printing press inspired fears that memory and wisdom would disappear as knowledge moved from oral tradition to page.

When television became widespread, literacy was declared doomed, with critics predicting that children would trade books for passive screen time and lose the stamina required for deep reading. When calculators entered classrooms, teachers warned that students would never truly understand mathematics again – that mental math would atrophy and foundational number sense would erode. Word processors were criticized for making revision "too easy," as though struggle were the only path to authenticity. Spellcheck was suspect, accused of weakening spelling ability and masking careless writing. When the internet arrived, knowledge was thought to be cheapened beyond repair, as information became instantly accessible and memorization seemed obsolete.

Over the centuries, each technological shift has provoked the same anxiety: that tools would replace thinking rather than extend it. Yet history suggests a different pattern – initial fear, gradual adaptation, and eventual integration into new forms of learning that expand, rather than erase, human capacity.

None of these concerns were entirely irrational. Each tool reshaped behavior, altering what was practiced and what was outsourced. Yet history reveals something equally consistent: education does not collapse – it adapts.

Artificial intelligence follows this same pattern, only faster and more visibly.

The fear surrounding AI in schools is not rooted merely in resistance to change. It emerges because educators care deeply about learning, fairness, and integrity. It also arises because the pace of technological innovation often outstrips the systems designed to support educators.

Humans naturally gravitate toward efficiency, and AI dramatically accelerates that instinct. At the same time, it exposes gaps between students' technological fluency and teachers' formal training, revealing weaknesses in preparation, professional development, and systemic support.

Few educators are as immersed in digital culture as their students. The rapid spread of AI trends – memes, chatbots, automated content creation – demonstrates how quickly these tools permeate learning spaces.

Teachers are left not resistant, but responsible. They must respond, often without guidance, policy clarity, or time to experiment thoughtfully.

Panic often emerges when responsibility arrives before understanding.

The panic cycle in education follows a familiar arc. First comes alarm: headlines predict the death of homework, writing, and authentic assessment. Next comes control: bans are issued, surveillance software is adopted, and rigid policies proliferate. Finally comes exhaustion, as teachers discover that enforcing these controls is unsustainable and may be counterproductive.

This pattern is not a failure of teachers. It is a systemic response that occurs when innovation collides with accountability structures built for a different time. "Education systems were designed for an earlier era and often do not reflect what we now know about how learning and development occur" (Darling-Hammond et al., 2020, p. 98). When structures lag behind science, alarm and overcorrection often follow.

When COVID closed school buildings, urgency arrived faster than clarity. As a principal in a residential incarcerated school setting, I was told we would "move to remote learning." The phrase sounded clean. The implementation was anything but.

Our students did not have computers in their living units. Technology access was controlled, limited, supervised. We did not have a ready-made online curriculum. Many of our teachers had never taught on Zoom. Some had never used it at all. Yet the expectation was immediate continuity.

So we improvised.

Teachers taught from home, logging into Zoom sessions students could only access if facility staff escorted them to designated spaces. Facility staffing shortages meant schedules shifted unpredictably. Some days students logged in late. Some days they did not log in at all. On other days, security priorities understandably overrode instructional ones. There were consecutive weeks when certain students were not permitted to access education in this format due to facility restrictions. We documented engagement. We worked to maintain standards. We tried to keep learning moving.

Inside faculty meetings, I watched urgency compress nuance. Teachers felt exposed – not because they lacked commitment, but because they lacked infrastructure. Those who were more comfortable with technology moved ahead quickly; others felt their competence shrink in real time. Equity, which we had worked so carefully to cultivate, became uneven overnight.

I remember sitting at my desk after one particularly chaotic day, staring at attendance data that looked more like evidence of disruption than learning. I felt the pressure to solve what was, in truth, larger than any single decision. The system demanded immediacy. The reality required patience, staffing, hardware, training, and grace.

That season revealed how quickly institutions default to acceleration under stress. We did not lack dedication. We lacked stability. And in the absence of stability, urgency filled the gap.

At no point in this panic cycle did understanding lead the conversation.

My friend Brian had a phrase for moments when well-intended efforts collapse into confusion: "What a soup sandwich." This names the reality that when too many reactive decisions stack upon one another, coherence dissolves. Something once planned to succeed results in a mess.

Education panic often resembles this. Policies layer onto policies. Tools are added without clarity. Expectations multiply without alignment. Each response makes sense in isolation, yet together they create a system that is difficult to navigate and nearly impossible to sustain. What emerges is not improved learning, but a tangle of contradictions that leaves teachers exhausted and students disengaged. Panic in education grows when fear overtakes thoughtful design.

Illustration in the Classroom: Elementary Level

Several essays arrive unusually polished.

Sentences are smooth. Vocabulary feels elevated. A few paragraphs share similar phrasing. Other students submit handwritten drafts marked with erasures, arrows, and revisions. The contrast is visible before a single question is asked.

The moment invites urgency. Is this misuse? Is this dishonesty? Should tools be restricted? The pressure to respond quickly – to protect standards, to assert control – rises almost immediately. Cautiously, the teacher waits.

Instead of beginning with enforcement, she begins with comparison. Two anonymous paragraphs are displayed side by side: one drafted independently, one produced with AI assistance. The class is asked not who wrote them, but what they notice. Which paragraph includes specific sensory detail? Which one feels general? Where does the writing sound lived-in? Where does it sound assembled?

Students begin identifying differences without accusation. The polished paragraph is coherent but broad. The handwritten draft includes concrete moments – the smell of rain, the scrape of a chair across the floor. The lesson shifts from originality policing to voice development.

Students are invited to revise both versions. They circle vague phrases and replace them with specific details. They annotate where thinking feels assumed rather than earned. AI is repositioned – not as forbidden, not as flawless – but as incomplete without human experience.

The classroom climate changes subtly. Instead of whispering about detection, students begin asking about purpose. When does a tool help? When does it flatten voice? What must remain ours?

The essays become less uniform – and more distinct.

The panic that might have hardened into surveillance becomes a study in craft. Standards remain intact. But they are upheld through design, not fear.

Illustration in the Classroom: Middle School Level

Lab reports are submitted.

Some are polished, comprehensive, and confident. Others are tentative – data carefully recorded, analysis still forming. The variation is sharp.

Questions arise quickly. Is the presentation masking misunderstanding? Is control slipping?

Students feel the shift before it is announced. Monitoring tightens. Reminders about integrity become sharper. The room edges toward the familiar second phase of panic: control.

The teacher interrupts that trajectory.

During the next lab, students bring three artifacts to the table: their raw data, their written explanation, and an AI-generated explanation of the same reaction. They are not asked to confess. They are asked to compare.

Where does your data diverge from the generalized explanation? Which variables did your trial include that the model ignored? Where does the clean explanation oversimplify what actually occurred? The conversation redirects attention from authorship to accuracy.

Students begin noticing discrepancies. Reaction times varied more than predicted. Color changes occurred unevenly. The generalized explanation assumes ideal conditions that were not present in the classroom experiment.

Reports are revised – not because enforcement demanded it, but because the evidence does. The tone in the room shifts from guarded to analytical.

The question is no longer "Did you use AI?" It becomes "Does this explanation hold up against your data?" Control yields to inquiry.

Standards are preserved not by tightening restrictions, but by strengthening reasoning.

Illustration in the Classroom: High School Level

Essays on *Macbeth* arrive. The prose is sophisticated.

Citations are correct. The arguments are coherent and confident.

In discussion, when asked to defend a claim about inevitability and moral decay, a student hesitates. The essay asserted certainty. The dialogue introduces complexity. Which line proves inevitability? Could ambition have been redirected? Does the text allow ambiguity? The fluency of the written page does not fully transfer to live defense. The tension in the room is immediate. Academic integrity, rigor, authorship – the conversation could turn quickly toward enforcement.

Instead, the teacher redesigns the next lesson.

AI-generated interpretations are projected alongside student work. The class annotates both; looking for generic reasoning, claims that lack textual specificity, arguments with flattened ambiguity, overstated certainty.

Students mark patterns. They identify what feels plausible but underdeveloped. They begin noticing where interpretation requires more than polished language.

Then the structure shifts again. Students must defend one claim orally, respond to counterargument, and revise their position publicly. Written essays are followed by live reasoning sessions. Revision is built into grading criteria – not as penalty, but as expectation.

Over time, the classroom culture changes. Students begin using AI selectively – consulting it for brainstorming or structural support – but grounding interpretation in textual evidence and personal reasoning. They anticipate challenges. They annotate where their thinking evolved. They treat revision as intellectual maturity.

The panic that might have narrowed into stricter surveillance becomes a lesson in discernment. The classroom does not eliminate tools. It situates them. Judgment, not prohibition, becomes the center of rigor.

Artificial intelligence use triggers panic not because of inherent dishonest use, but because teachers are asked to uphold standards while navigating tools for which they received little preparation. Alarm leads to control. Control leads to exhaustion.

But when educators recognize this cycle, they can interrupt it.

Instead of reacting to outputs, they can redesign experiences. Instead of policing tools, they can cultivate judgment. Instead of escalating enforcement, they can clarify expectations about reasoning, voice, and understanding.

Recognizing this predictable arc: initial alarm, attempts at control, and eventual fatigue, allows educators to shift from reaction to reflection. When teachers guide students to use AI as a support for thinking – rather than a substitute for it – panic gives way to purpose.

The issue is not whether students will use AI. The issue is whether we respond with fear or with design. And that distinction determines whether innovation destabilizes learning – or strengthens it.

History reminds us that when fear drives response, clarity is often the first casualty.

One example of this may be found during the COVID-19 pandemic, when many schools closed abruptly, leaving teachers without training, curricular adjustments, or sustained professional support. Yet they were still expected to deliver meaningful instruction, maintain engagement, and meet learning outcomes amid chaos. Many improvised on the fly, relying on tools and practices they had little time to master. Artificial intelligence accelerates a similar dynamic: the tool is present, students are using it, but the system has not equipped teachers with strategies to harness it productively rather than punitively.

This is a systemic pattern that reappears when innovation collides with accountability structures shaped by another time. When this collision is amplified by the pressures of big business, reactionary adoption cycles accelerate – cycles most educators recognize not from theory, but from experience.

It is within this churn that frustration often surfaces. Parents and teachers alike may feel unsettled when familiar methods give way to unfamiliar strategies. The tension between "They're not teaching math the way I learned it" and the reality of contemporary learning is deeply human. Bridging this gap requires patience, curiosity, and adaptability – not nostalgia, and not an overreliance on AI. Students need guidance that honors both evolving pedagogy and the lived realities of their learning environments. Few examples capture this cycle of urgency, influence, and overcorrection more clearly than the evolution of literacy reform.

If you have not yet listened to the podcast *Sold a Story*, hosted by Emily Hanford (2022), it offers a revealing lens into curriculum adoptions. Hanford investigates the ebb and flow of literacy movements, the influence of big business, and the systemic pressures surrounding how students are taught to read. Her reporting illustrates how educational systems prioritize tools over teacher expertise. Understanding becomes secondary to adoption. Teachers are provided with packaged solutions rather than cultivated as professionals capable of discerning what works and why.

When I listen to her podcast, I am reminded of countless requests – and at times administrator mandates – for "plug-and-play" resources. Tools that promise immediate results, materials designed for rapid implementation. Programs marketed and purchased to "fix" whatever problem has surfaced in what Safir and Watson (2021) describe as satellite-level data – the level farthest from students, composed of test scores, compliance metrics, and performance indicators. In these moments, response becomes transactional rather than relational, driven by what is easiest to measure rather than what is closest to learning.

Satellite data can signal that something is wrong, but it cannot explain why. Yet too often, solutions are selected at the same altitude as the problem, privileging efficiency over understanding. Rarely are we asked what would deepen teachers' craft, strengthen instructional judgment, or cultivate the relational expertise that cannot be packaged.

This tension between quick-fix tools and professional expertise reveals how systemic pressures shape classroom practice, subtly incentivizing speed and visibility over thoughtful pedagogy. School budgets reinforce this pattern, with line items earmarked for curriculum adoptions that imply purchase as proof of action, regardless of coherence or contextual fit.

Acquiring a new program can create the reassuring sense that we are addressing a concern – that we are accountable and responsive. But when investment remains at the satellite level rather than moving toward what Safir and Watson (2021) call street-level data – the lived experiences, voices, and daily realities of students and teachers – the response risks being efficient yet incomplete, visible yet disconnected from the deeper work of learning.

This preference for ready-made solutions does not emerge in isolation; it is sustained by systemic incentives that reward speed, visibility, and measurable output over professional depth.

The recurring cycle of alarm, control, and exhaustion is reinforced by these incentives. Big business markets new curricula, platforms, and tools as solutions, often promising efficiency, compliance, or standardized success. Administrators and teachers, constrained by accountability metrics tied to visible outputs, adopt tools rapidly – sometimes without adequate training or pedagogical reflection. The result is predictable: teachers are expected to implement without mastering craft, and students are expected to comply without developing judgment.

Even when educators aim to cultivate inquiry, problem-solving, and adaptability, systemic pressures can reward correct answers over deep thinking.

When we focus exclusively on equipping teachers with the newest curriculum, platform, or strategy, we inadvertently undermine their ability to master the art of teaching. A balance is needed.

Artificial intelligence makes visible a truth education has long avoided: many school practices depend on the scarcity of information rather than the cultivation of reasoning. When scarcity disappears – when answers are instantly available – practices built around retrieval tend to wobble.

Rather than asking how to eliminate AI, a more productive question emerges: What are we actually trying to protect? If the answer is learning, then fear-driven policies rarely serve us well. Banning tools does not teach discernment. Surveillance does not build integrity. Control does not foster curiosity.

Teachers understand this intuitively. Most resistance to AI is not about the tool itself; it is about the absence of guidance. Educators are being asked to police something they have not been invited to understand, while being held accountable for outcomes beyond their direct control.

The panic cycle thrives in that gap.

This pattern is not new. Teachers are routinely asked to adapt on the fly to absenteeism, background knowledge gaps, shifting standards, and changing assessments – often without sustained support. Yet evaluations and funding remain tied to grade-level outputs. The pressure to produce visible results, regardless of context, fuels reactive decision-making.

This can feel destabilizing. But it is also an opportunity.

Breaking the panic cycle requires a shift – from reaction to reflection, from enforcement to education, from asking "How do we stop this?" to asking "What does this reveal?" When educators move beyond panic, they regain agency. They stop competing with technology and start redesigning learning. They focus less on policing tools and more on teaching students how to use them responsibly, think critically, and cultivate judgment.

AI reveals which assignments measure thinking and which measure compliance. It exposes where grading rewards output over process. It highlights how rarely students are required to explain, defend, and revise ideas in sustained ways.

Panic narrows vision; perspective expands it. AI is not the first disruption education has faced, and it will not likely be the last. The question is not whether change is coming, but whether educators meet it with anxiety or intention. History – and decades of innovation – suggest that learning survives best when teachers are trusted to lead.

When schools prioritize clarity, trust, and professional judgment, AI becomes a tool for reflection rather than a trigger for fear. It becomes a catalyst for redesign rather than a threat to integrity. And in that shift – from panic to purpose – education strengthens rather than fractures.

Invitations for Preservice Teachers

Entering the profession during a time of rapid technological change can feel unnerving.

Tools evolve faster than policies. Expectations shift before training catches up. And when responsibility arrives before understanding, panic can fill the gap where clarity should live. You can reframe that experience: panic is not evidence of incompetence – it is a predictable human response when systems collide with uncertainty.

You are invited to consider: When uncertainty surfaces, what posture will you adopt? Will you rush toward control – or pause toward clarity? What habits will anchor you when expectations feel unstable?

It would be difficult to name a generation of teachers who entered a truly stable profession. Calculators, the internet, standardized testing reforms, shifting literacy movements – each era has confronted tools that felt disruptive and overwhelming. What distinguishes strong educators is not their ability to eliminate uncertainty, but their capacity to respond to it thoughtfully.

You are not expected to have all the answers. What matters is your willingness to pause rather than react. To design experiences that prioritize reasoning, dialogue, and revision rather than polished outputs. When you focus on cultivating thinking rather than reacting, you model steadiness in a system that often rushes toward control.

When you meet new tools with curiosity rather than fear, you teach students how to do the same. In doing so, you help them learn not only content, but how to think calmly and critically in uncertain times. That steadiness is not accidental – it is leadership in the classroom.

Invitations for Administrators

The panic cycle is not primarily a classroom problem.

It is a leadership one.

When anxiety drives policy, schools move quickly toward surveillance, restriction, and compliance. These responses may feel decisive, but they often erode trust and narrow instructional practice. Interrupt that pattern.

You are invited to consider: What emotional tone does your leadership set when uncertainty enters the system? When innovation emerges, do you amplify urgency – or cultivate clarity? Are your responses stabilizing, or reactive?

When innovation emerges, leaders set the emotional tone. If urgency dominates, teachers feel pressured to control rather than design. If clarity leads, teachers feel empowered to reflect and adapt. That difference shapes school culture.

The Lippitt-Knoster Model for Complex Change offers a helpful lens here. The model reminds us that sustainable change requires five elements: vision, skills, incentives, resources, and an action plan (Knoster et al., 2000). When any one of these is missing, predictable dysfunction occurs – confusion without vision, anxiety without skills, resistance without incentives, frustration without resources, or false starts without a clear plan. Much of the AI panic in schools reflects precisely this breakdown.

Teachers are asked to respond to AI without a clearly articulated instructional vision, without structured professional learning to build competence, and without protected time to experiment and refine practice. Anxiety, then, is not a personal failing; it is a systemic signal that essential components of complex change have not yet been aligned.

Leadership in an AI era requires restraint as much as action. It requires articulating a clear vision for how AI aligns with learning goals. It requires investing in skill development before imposing policy enforcement. It requires aligning incentives so that depth of reasoning – not mere compliance – is rewarded. And it requires providing practical structures that allow teachers to experiment safely and thoughtfully.

Administrators

The most effective leaders recognize that reaction compounds confusion. Policies layered onto policies create "soup sandwich" systems – complex, contradictory, and exhausting. But when administrators ground decisions in how learning actually works, they stabilize the environment. They foster and protect teachers' professional judgment. They ensure that innovation serves pedagogy rather than dictating it.

You are your teachers' strongest advocate. When you respond to change with vision instead of urgency, you reduce panic and create coherence. When you align skills, resources, and expectations before demanding implementation, you transform anxiety into momentum. In this way, leadership becomes not a source of pressure, but a stabilizing force that keeps learning human – even as tools evolve.

Invitations for Homeschool Parents

Homeschool families may sit outside institutional systems,

but they are not outside the conversation. Even at a distance, you may feel the ripple effects of the educational panic cycle. Discussions about AI often carry tones of alarm or inevitability. Even without district mandates, cultural pressure can create anxiety. For some homeschool families, state accountability requirements – assessment, documentation, portfolio reviews, or shifting regulations – may also evolve in response to public school policy, introducing indirect pressures that mirror broader debates.

You are invited to consider: What tone shapes your home when new technologies emerge? Do headlines dictate your response, or does reflection? When anxiety surfaces, do you absorb it – or translate it into thoughtful conversation?

You are invited to consider how your calm becomes curriculum – how responding thoughtfully, rather than reacting defensively, models discernment.

Homeschooling offers flexibility that many institutional settings cannot. You can pause, experiment, and reflect. This freedom allows you to approach AI not as a threat to eliminate, but as a tool to evaluate. When you model calm decision-making and intentional tool use, students learn that new tools can be examined, questioned, and integrated thoughtfully. Agency is preserved not by rejecting innovation, but by engaging it with judgment.

In a culture that equates urgency with importance, your steadiness becomes instruction. Learning is not defined by speed, compliance, or novelty – it is defined by reflection, judgment, and growth. That lesson endures long after specific tools fade.

Reflection Questions

Chapter 7: The Panic Cycle in Education

1. Is your response to this technological shift rooted in evidence – or in anxiety?

2. What institutional pressure are amplifying panic I your context – and who benefits from that amplification?

3. How does the current moment resemble past educational 'crisis' – and how did those cycles resolve?

4. When you first chose this profession, what problem were you hoping to solve – and has that problem actually changed?

5. If you chose steadiness instead of speed in this moment, what would that look like in your classroom or leadership?

Chapter 8: When Students Use AI Before We're Ready

When panic begins to subside, a different question emerges. If alarm and control are not sustainable responses, what is actually happening? Beyond the headlines and policy debates, beyond the surveillance software and staff meetings, the real shift unfolds quietly – assignment by assignment, interaction by interaction. The panic cycle describes the institutional reaction. What follows requires closer attention to student behavior itself.

The signs are quiet at first – a polished sentence here, an explanation offered without hesitation, work that is technically correct but strangely detached. Teachers often sense something is off long before there is proof.

The signal is rarely error but one of absence. The reasoning that normally unfolds across drafts disappears. The small hesitations that accompany genuine thinking are gone. The work arrives complete, but the intellectual fingerprints are missing.

This is what makes the shift difficult to diagnose. Nothing appears obviously wrong. The answer is correct. The grammar is clean. The structure is organized. What is missing is the visible path that leads to understanding.

By the time schools begin debating policies, many students have already begun experimenting. This is not a failure of discipline. It is a predictable response to access. Students use the tools available to them. They copied homework before calculators existed. They searched the internet before teachers learned how to cite it. They used spellcheck long before anyone decided whether it was "allowed."

Each technological shift follows a similar pattern: quiet experimentation first, informal sharing among students next, institutional awareness later. AI is simply the newest expression of an old truth: learners adapt faster than systems. The mistake is assuming that early use is the same as misuse.

Illustration in the Classroom: Elementary Level

Students are assigned a book report.

One student opens an AI tool at home and types: "Write a fourth-grade book report about *Charlotte's Web*." Within seconds, a well-structured report appears. It includes clear and coherent report elements. The grammar is strong. The transitions are smooth. The student turns it in.

The teacher recognizes the structure immediately – not because it is wrong, but because it is symmetrical. Balanced. Predictable. The organization is clean, but the voice is distant. Rather than framing the moment as misconduct, she reframes it as mentorship.

The next day, she projects an anonymized AI-generated book report and asks the class to analyze its structure. Where is the thesis located? How are the paragraphs organized? What makes the transitions clear?

Students identify patterns quickly. The report follows a reliable formula: summary first, character description second, theme third. "Now," the teacher says, "what does it *not* include?" The room grows quieter. Students notice the absence of personal reaction. There is no moment of surprise. No line that lingered. No description of how the story felt.

The AI report becomes a scaffold. Students outline its structure in their notebooks, then use that framework to rebuild their own drafts – this time inserting lived detail. They add favorite scenes. They include confusion and delight. They write sentences that bend slightly out of symmetry because they carry perspective. The tool remains useful. It modeled organization clearly and efficiently. But structure is not substance.

In this way AI has become helpful for framing ideas, insufficient for filling them. Students begin distinguishing between form and voice, between organization and meaning. The lesson is not that AI cannot write. The lesson is that writing is more than structure.

Illustration in the Classroom: Middle School Level

Students are examining a recent news article about a controversial community decision. Opinions are already forming – at home, in hallways, online.

One student generates an AI summary of the article. It is concise. Balanced. Neutral in tone. It lists key stakeholders and outlines the decision without emotional language. The summary appears fair. The teacher projects both the original article and the AI-generated summary side by side. "What changed?" she asks.

Students note that the summary removes charged language. It compresses quotes. It simplifies context. "Is that better?" the teacher asks. The room hesitates. They begin rereading the original article closely. They circle phrases that frame one group as "concerned parents" and another as "activists." They identify statistics that appear persuasive but lack context. They notice that the AI summary flattened those distinctions. It preserved structure – but erased tone. The summary is not incorrect. It is incomplete.

Students begin asking sharper questions: Who benefits from neutrality? What happens when complexity is compressed? Is omission a form of bias?

The AI output becomes a case study – not in failure, but in framing. Students revise the summary, reintroducing nuance. They specify whose voices were amplified and whose were minimized. They annotate where perspective shapes interpretation. The technology remains present, but it has shifted from authority to artifact.

In middle school – where identity and autonomy are forming – this distinction matters. Students are learning that information ecosystems reward speed and clarity. But clarity can conceal complexity.

AI succeeded in summarizing.

The work of the classroom is to interrogate what summary leaves behind.

Illustration in the Classroom: High School Level

Students complete a lab investigating reaction rates under varying temperatures. One student drafts her analysis using AI as a starting point. The report is well organized. Terminology is accurate. The explanation of collision theory is precise. The writing is strong.

When asked how temperature influenced molecular motion in her specific trial, she refers first to her own data. She identifies where her reaction rate deviated slightly from predicted values. Then she explains how she used AI to clarify phrasing around activation energy. "The wording helped," she says. "But I had to adjust it to fit our results." The teacher presses gently. "What did the tool miss?" "It assumed ideal conditions," she replies. "Our thermometer wasn't perfectly calibrated." The AI explanation provided technical clarity. It did not anticipate the imperfections of her experiment.

During revision, she keeps the structured explanation of collision theory but adds a paragraph on uncertainty – how slight temperature variation influenced particle speed differently than expected. She annotates the draft to indicate where AI assisted with phrasing and where her interpretation guided the analysis. The tool functioned as assistant – not author.

Over time, her process stabilizes. She drafts interpretations independently before consulting AI for technical refinement. She uses it to check terminology, not generate reasoning. The workflow becomes layered: Observation → Interpretation → Clarification → Revision.

AI supports precision. It does not replace conceptual ownership. In this classroom, success is not measured by avoidance of technology. It is measured by sequencing it properly.

Artificial intelligence performed well. It organized language and clarified theory. Human judgment determined how that theory applied to imperfect reality.

The distinction is subtle – but foundational.

Students often turn to AI not to cheat, but to cope. They use it to manage cognitive load, reduce frustration, meet deadlines, and navigate expectations. In many cases, AI becomes a mechanism for survival in systems that reward polished output.

The presence of AI does not automatically signal misconduct. It signals access.

The deeper question, then, is not whether students are using AI, but why – and what that use reveals about our assignments, pacing, and expectations. When we treat every instance as misuse, we miss the opportunity to teach discernment. When we slow down long enough to examine the pattern, we see something clearer:

Learners adapt quickly. Systems adapt slowly.

To guide adaptation, we must first understand the conditions that make it inevitable: "Successful adaptation to emerging technologies requires that organizations first establish shared understanding of both the capacities and limitations of those technologies, and then align structures, norms, and practices accordingly" (Van Allen & Freedberg, 2024, p. 53).

AI is already woven into students' daily lives, so it becomes the tool they reach for when challenges surface or support feels thin. It is not distant technology housed in a lab; it lives in the same devices they use to connect, play, and explore – where school demands and personal curiosity meet. Its presence is constant: before school, after school, and in the quiet moments when deadlines press. For students navigating high expectations and uneven support, AI is not merely convenient – it can feel essential. What feels indispensable to students becomes instructional territory for teachers.

Teachers who rise to this moment do more than enforce rules; they transform it into an opportunity for authentic learning. They shift the conversation from detection to discernment, intentionally making thinking visible in ways that cannot be replicated by polished output alone.

Drawing on principles articulated in *Building Thinking Classrooms in Mathematics* (Liljedahl, 2021) and *Making Thinking Visible* (Ritchhart et al., 2011), teachers can design tasks that require students to externalize reasoning, justify decisions, revise publicly, and engage in collaborative sense-making. They structure learning so that cognition is observable – through whiteboard problem-solving, visible drafts, thinking routines, and dialogic reflection – rather than hidden behind completed products. In doing so, they ensure that understanding cannot be outsourced, because the process of thinking itself becomes the evidence.

Ownership of knowledge – demonstrated through explanation, transfer, and intellectual risk-taking – replaces polished output as the true measure of success.

When teachers frame AI as a thinking tool rather than a cheating tool, they model judgment, ethical reasoning, and responsibility – the very capacities they hope students will internalize. Students often respond with honesty and curiosity when invited into dialogue about AI's role. They are not seeking loopholes as much as they are seeking clarity.

In this context, AI is not the problem students are bringing into the classroom. It is a mirror. It reflects pressures, anxieties, and systems that prioritize product over process. Students turn to it when learning environments have not fully scaffolded reflection, reasoning, or persistence. Responding with punishment alone addresses none of these underlying needs.

Teachers who choose preparation over policing, guidance over restriction, and reflection over suspicion empower students to navigate a world in which tools will always evolve faster than rules.

This moment is also an invitation for educators. It calls teachers to explore AI thoughtfully, to understand both its affordances and its limits, and to design learning experiences that leverage its potential to cultivate curiosity, agency, and resilience. In doing so, teachers reclaim their role as architects of understanding rather than enforcers of compliance. They model how to approach new tools with discernment, courage, and intention.

When students use AI before we feel ready, it is not a crisis to contain. It is a signal – to teach more explicitly, to design more intentionally, and to guide with humanity as the steady constant amid accelerating change.

Invitations for Preservice Teachers

Your task is not to outrun AI – it is to understand what student behavior is trying to tell you. Behavior is communication.

Students can encounter new tools before schools have fully decided how to respond to them. This is the natural intersection of curiosity, pressure, and access. Consider interpreting early AI use not as rebellion, but instead as communication. When students turn to AI, they are often signaling something: uncertainty about expectations, anxiety about performance, or confusion about how to begin.

You are invited to consider: What is this behavior revealing about your design? Where might expectations be unclear, cognitive load too heavy, or purpose insufficiently articulated? Before correcting the student, can you diagnose the system?

When students use AI prematurely, treat the behavior as instructional data. What does it reveal about task design, cognitive load, or clarity of purpose? Your discernment lies in diagnosing systems before disciplining individuals. Your role is not to outrun technology. It is to slow thinking down. To ask, "How did you arrive at that conclusion?" "What part of this is yours?" "Where does your reasoning deepen – or thin?" When you design assignments that require explanation, defense, and revision, you shift the focus from product to ownership. You make understanding visible. You model that AI can assist brainstorming or organization, but it cannot substitute for analysis or judgment. When you respond with curiosity instead of suspicion, you communicate trust – and trust invites honesty. Students are far more likely to engage reflectively when they are invited into dialogue rather than pushed into secrecy.

In a profession where tools can arrive before policies, your steadiness becomes instruction. You teach students that discernment is learned, not automated. That ethical reasoning requires conversation. That ownership of ideas matters more than polish. And that understanding cannot be outsourced, even when language can.

Invitations for Administrators

When students adopt tools before policies are finalized, the instinct to respond quickly is understandable. Accountability structures demand visible action. Families expect clarity. Teachers look for guidance. Yet speed is not the same as wisdom.

AI use is often a system signal. It reveals where expectations emphasize output over process, where pacing outstrips mastery, or where students lack scaffolding for reasoning. If leadership interprets every instance as misconduct, the response narrows to surveillance and enforcement. But if it is interpreted as feedback, a different path emerges.

You are invited to consider: What is this behavior teaching you about your design? Are students responding to pressure you unintentionally created? When you see misuse, do you move first toward restriction – or reflection?

Resist surveillance-based solutions. Early AI misuse often signals design misalignment, not moral collapse. Lead investigations into structural incentives rather than defaulting to enforcement.

Leadership in this moment requires reframing. Students are navigating environments that reward polished work under time pressure. Without explicit instruction on discernment, they will naturally leverage tools to meet those expectations. Rather than intensifying control, administrators can invest in professional learning that equips teachers to redesign tasks – prioritizing explanation, defense, dialogue, and revision. Policies are necessary, but they must be coupled with pedagogy. Surveillance without instructional redesign breeds distrust. Support paired with clarity builds capacity. When administrators create space for teachers to experiment thoughtfully, collaborate, and reflect on AI's role, innovation becomes teachable rather than threatening. Leaders who respond with support instead of suspicion shift the narrative. They move schools from detection to discernment. From compliance to understanding. In doing so, they ensure that emerging tools serve learning rather than destabilize it.

Invitations for Homeschool Parents

AI often enters learning quietly – through curiosity, convenience, or the desire to get it right.

A student may use it to clarify a math explanation, summarize a dense article, or refine a draft. Early use does not necessarily signal dependence. It signals exploration.

You are invited to consider how you will shape that exploration before it becomes habit. You are uniquely positioned to make that exploration visible and reflective. You have the flexibility to pause and ask questions that deepen awareness: "What did the tool help you understand?" "Where did it simply give you words?" "What do you still need to figure out yourself?" These conversations turn passive consumption into active reasoning.

You are invited to consider: When AI appears in your student's work, do you move quickly past it – or linger with it? Are you modeling curiosity about the process, not just the product? How will you help your student distinguish between assistance and authorship?

You can model balance. AI can support organization, clarify vocabulary, or generate examples. But thinking – the slow work of wrestling with ideas – remains human. When you invite your student to explain concepts in their own words, apply ideas in new contexts, or critique what the tool produced, you reinforce intellectual ownership.

In guiding these reflections, you teach more than digital literacy. You teach discernment. You show your student how to remain a thinker in the presence of powerful tools. And in doing so, you ensure that curiosity leads not to dependence, but to deeper understanding.

Reflection Questions

Chapter 8: When Students Use AI Before We're Ready

1. If students are using AI, what might that reveal about the tasks we are assigning?

2. Which of your assignments can be completed convincingly without genuine understanding?

3. What kind of learner are you truing to cultivate when no one is watching?

4. Are your AI policies designed primarily to prevent misuse – or to develop judgment?

5. What would it mean to be ready – not to stop AI use, but to design around it intentionally?

PART III — THE ART OF TEACHING IN AN AI WORLD

Chapter 9: Teaching as Design, Not Delivery

Students will adapt to the tools available to them. The question is not whether AI will be present in learning, but how learning itself is structured in response. When early AI use becomes guidance instead of panic, another realization emerges – technology does not redefine teaching. It exposes what teaching has always been.

For much of modern schooling, teaching has been described as delivery. Curriculum is delivered. Instruction is delivered. Content is delivered. The language suggests a transaction: information moves from expert to learner, and learning occurs if the transfer is successful.

In practice, this metaphor often materializes as "stand-and-deliver" instruction – the teacher positioned at the front of the room, speaking while students receive, as though knowledge travels in a single direction. The metaphor implies efficiency. It assumes that knowledge is a package – intact and stable – waiting to be transported. In this framing, teaching resembles shipping: once the package is assembled, the task is simply to move it from one location to another. If delivery is successful, the job is done.

But learning has never functioned like freight.

A more accurate metaphor is architectural. Teaching is not shipping a package; it is designing a building. Delivery assumes something is already complete and merely needs transport. Architecture assumes something must be intentionally constructed for people to inhabit.

AI can ship information. It can generate explanations and distribute answers at scale. Teachers, however, design structures students can think inside. They create environments where ideas connect, where misconceptions surface, where questions stretch understanding, and where learners move through intellectual space with support and intention.

In education, I have often heard, and said myself, we are "building the plane as we fly it" to capture the process of implementation without adequate planning, vetting, time, and consideration. I have used this phrase when reacting to a need, with the best available option, with all fingers crossed, and a smile on my face, hoping it will work out and hoping implementation will land softly. The phrase captures the reality that teaching is dynamic and adaptive; we refine instruction even as learning unfolds. But even that metaphor implies motion without structure. Architecture deepens it. We are not merely assembling parts midair; we are intentionally designing spaces sturdy enough to hold uncertainty, revision, and growth. The work is not frantic construction – it is responsive design.

For decades, the delivery model appeared sufficient because information was relatively scarce, assessments rewarded correct answers, and classrooms were structured around compliance and efficiency. When knowledge was harder to access, transferring it seemed like the central task. Learning is not transactional. It is relational, interactive, and emergent.

If teaching were merely delivery, AI would already be the superior teacher. It does not tire. It generates explanations instantly. It adapts tone and complexity in seconds. It produces fluent output without hesitation. By the logic of delivery, it should win.

And yet something essential is missing – not because the technology is deficient, but because learning is not a shipment to be transferred or received.

Learning is an experience.

It unfolds in the dynamic space between teacher, content, and learner. It requires missteps, feedback, recalibration, and trust. AI can simulate structure and supply answers, but it cannot cultivate curiosity, scaffold reasoning, normalize struggle, or restore confidence after failure. Fluency does not equal understanding. Access does not equal insight. Information does not equal transformation.

Teachers do not deliver knowledge; they design the conditions under which understanding can emerge.

I once facilitated an art project with students in a residential incarcerated setting that, at the time, felt powerful. We were discussing life beyond incarceration – what it would take to rebuild, to persist, to stay free. I handed out dry erase markers and invited students to write on their arms the qualities they believed would sustain them on the outside. Words appeared slowly:

discipline, patience, self-control, faith, consistency, forgiveness. Some students layered their forearms with phrases. Others wrote one word in careful block letters. There was laughter. There was pride. There were quiet moments of resolve.

It was inspirational. It was meaningful. It was memorable.

It was also poorly designed.

The activity generated emotion and reflection, but it did not anchor itself to transferable learning. We did not analyze the conditions that make those traits sustainable. We did not connect perseverance to specific economic systems, social networks, or cognitive habits. We did not examine recidivism data or study the psychology of habit formation. We did not build a structured bridge between aspiration and application.

The project created a moment. It did not construct a framework.

Since then, I have imagined how it could have been created differently. Students could have mapped each written trait to research on behavioral change or trauma recovery. We could have examined longitudinal data on reentry outcomes and identified which supports most strongly predict stability. They could have designed personal action plans grounded in evidence rather than sentiment. We might have paired the art with a writing component analyzing how identity narratives shape decision-making. We could have moved from inspiration to strategy.

The activity was not wrong. It was incomplete.

Design requires more than resonance. It requires intentional pathways from emotion to cognition, from expression to structure, from hope to habit. I learned that day that powerful experiences do not automatically become powerful learning. They must be built to endure.

That realization reshaped how I understood the work of teaching.

Teachers sequence experiences. They create cognitive tension and release. They anticipate misconceptions before they surface. They decide when to intervene and when to allow productive struggle. They shape emotional architecture and intellectual risk. Like architects designing spaces for human movement and interaction, teachers design environments for thinking.

This work requires judgment. Delivery does not.

When teachers shift from seeing themselves as deliverers to designers, everything changes. Lessons become pathways rather than scripts, questions become invitations rather than checkpoints, and assessments become feedback rather than judgment.

AI fits naturally into this model; not as a replacement for teaching, but as a material within.

Teachers can leverage AI to generate examples, suggest practice sequences, or provide alternative explanations, freeing cognitive bandwidth for coaching, interpreting reasoning, and cultivating understanding.

Designing learning in an AI-infused world demands new questions. Not, "What information must students receive?" but, "What thinking must students practice?" Not, "How do I prevent shortcuts?" but, "What experiences make shortcuts irrelevant?"

This reframing shifts attention from controlling tools to orchestrating experiences: spaces where students cannot outsource thinking, where they must explain, defend, revise, and apply ideas in novel contexts.

Thoughtfully designed learning experiences make thinking inescapable; such learning calls for collaboration, reflection, iteration, and judgment – capacities beyond AI's reach. Students learn most deeply when invited to engage, wrestle with ideas, and see their efforts mirrored in meaningful guidance and feedback. While AI can support practice, it cannot create the conditions in which these experiences naturally unfold.

A design mindset paired with a growth-oriented perspective honors the emotional architecture of learning by intentionally constructing environments where risk and reflection are safe. Learning spaces communicate who belongs and whether mistakes signal failure or development. "Young people are exquisitely sensitive to the messages environments send about who belongs, who can grow, and whether mistakes signal deficiency or development" (Yeager, 2022, p. 41). Growth, therefore, is not a matter of individual grit but of design. "A culture of growth...is about designing environments that normalize challenge, invite risk, and make learning from mistakes safe and expected" (Murphy, 2023, p. 24). When challenge is normalized and revision respected, the emotional architecture of learning supports genuine development.

Teachers plan for safety, risk-taking, and belonging. They design struggle as part of the process, normalize mistakes, and scaffold revision rather than punishment. They create cues – verbal, structural, and relational – that communicate, "Your growth is expected and supported."

In classrooms, emotional scaffolding is inseparable from cognitive development. Students think more deeply when they feel safe enough to risk being wrong. AI cannot perceive fear, validate effort, or repair trust; it cannot create the human conditions in which understanding flourishes.

Teachers who embrace design integrate emotional architecture with intellectual challenge, creating spaces where both risk and reflection are safe and expected. In these environments, students begin to see mistakes as opportunities rather than verdicts, persisting through difficulty and embodying a growth-oriented approach to learning. "Students' mindsets are shaped by the messages they receive from their learning environments about the meaning of difficulty and the potential for growth" (Yeager & Dweck, 2012, p. 303).

When teaching shifts from delivery to design, something else shifts as well: professional agency. Teachers no longer need to react defensively to technology or attempt to police every tool students use. Instead, they become intentional curators of learning, deciding where AI can support practice, spark discussion, or step aside entirely. The question is no longer, "How do we stop this?" and changes to "How do we design with this in mind?"

With that shift, the focus moves from enforcement to intention, from fear to possibility. Teachers can concentrate on the craft they excel at: creating environments where students grow cognitively, socially, and ethically. They become architects of learning, guiding students to think critically, reason independently, and navigate complexity, while leveraging AI to enhance – rather than replace – the experience.

Delivery ends when information arrives. Design begins when learning starts. Teaching is more than delivery because understanding does not emerge from exposure alone; it grows through guided struggle, reflection, and relational support.

And in embracing design, teachers reclaim the essence of their work: shaping not just what students know, but how they think, feel, and grow. AI may accelerate the pace of change, but it cannot replace the judgment, care, and artistry that define the art of teaching.

When teaching is understood as design rather than delivery, learning becomes lived, relational, and transformative. Even in settings where technology is limited or absent, the principles of design – intentional, relational, and student-centered – remain central to meaningful learning.

Tools may shift. Platforms may evolve. The architecture of learning does not.

My friend Brian had a way of naming misalignment with disarming clarity. When someone had every resource available yet still missed the point, he would say they "couldn't hit water if they fell out of a boat." The phrase is humorous, but the insight is precise. In education, access to tools, curricula, and platforms does not guarantee meaningful learning.

Abundance without intention is not an advantage.

When instruction is reduced to delivery – scripts followed without attention to context, learners, or purpose – even the best materials fail to land. Teaching as design demands something different: deliberate choices about what students will consider, struggle with, revise, and ultimately understand together. Without that intentional design, abundance becomes noise, and learning remains just out of reach.

This distinction between abundance and intention becomes unmistakably clear in settings where instructional design must stand on its own – without the illusion that technology, polish, or packaged curriculum can substitute for thoughtful teaching.

I once had the privilege of teaching in such a setting with almost no substantive technology. I taught in a residential incarcerated school where internet access was fully restricted and only the teacher workstation had limited connectivity. Students had access to desktop computers in the classroom, but even their most basic features had been removed. These restrictions were purposeful, designed to meet security requirements established by facility leadership.

In that environment, my challenge was clear: How do you bring research, creativity, and rich content to students when access is constrained? The answer was not more delivery – it was design. I brought in flash drives loaded with research articles and images so students could explore Native tribes while studying state history. I curated video clips that offered visual explanations when live demonstrations were not possible. Even with limited tools, students engaged meaningfully because the learning experience had been intentionally constructed.

In this classroom, access to technology was viewed as a privilege, sometimes even as reward. That reality sharpened my awareness of how essential access to information is – and how easily we take it for granted.

When design was thoughtful, scarcity did not prevent learning.

Yet the environment shaped more than access; it shaped perspective.

As a teacher in that setting, I was under constant surveillance. I never knew whether my classroom behavior was being observed live, recorded for later review, or simply monitored in passing. At times, a booming security voice would interrupt over the intercom to correct a student behavior I had missed – or was still deciding how to address. Occasionally, students would reference what I had eaten for lunch, revealing they could see the classroom from their living units when I was not teaching. The message was clear: observation was constant.

In addition to surveillance, my classroom included multiple adults. I was fortunate to work with an instructional aide – a paraprofessional – who supported instruction, assisted with classroom management, and collected data. Security guards were also present, often seated discreetly in corners, monitoring specific students' behavior. Their roles were distinct from mine, but their presence shaped the environment.

Those additional adults did more than support logistics – they influenced how I came to design instruction itself, shaping my understanding of what it meant to be observed in the art of teaching. They served multiple purposes, particularly in my early years of teaching, and their constant presence altered my perception of visibility, feedback, and performance.

Most educators will never experience that level of ongoing surveillance or the steady presence of non-instructional staff. Many teach alone, responsible for delivering content, maintaining engagement, ensuring safety, and guiding learning without additional adult support. In any setting, however, one truth remains: the teacher is only one human being. That person has blind spots, biases, strengths, and limitations. Inevitably they will catch some nonverbal cues and miss others. It is tempting to imagine that artificial intelligence might fill those gaps.

However, AI does not function as surveillance or instructional partnership in the same way humans do. It cannot read tone shifts, interpret body language, or understand the layered context students bring into a room in the same way humans can. It cannot whisper a well-timed question that redirects instruction. It cannot sense when a teacher is drifting from intention.

One of the most valuable systems my instructional aide and I developed was designed not for students – but for me.

There was one student who frequently initiated thoughtful conversations about social justice. He asked about youth incarceration, socioeconomic inequities, food access, and systemic barriers – topics deeply connected to his lived experience. I appreciated those conversations. I welcomed the opportunity to stretch thinking, pose questions, and explore societal structures together. These dialogues were rich, human, and meaningfully connected to learning. They built trust in our classroom, and I valued them.

There was just one problem: they consistently occurred during independent math time.

This student perceptively discovered that my interest was an easy pivot point. With a well-placed question, he could redirect the entire classroom from math practice to social justice. It was clever. It was predictable. It was highly effective at derailing math. And I did not catch that it was becoming routine.

While, I did not notice the pattern at first, my instructional aide did. She gently pointed out that, for the third time that week, independent math work had dissolved into extended discussion. She observed that once these topics surfaced, most students disengaged from math entirely, and I supported the derailment by encouraging the discussion. Her observations were accurate.

Initially, I resisted. It felt as though acknowledging the pattern meant admitting I had misplaced instructional priorities. That somehow the value I prescribed to our conversations was lessoned by the timing of them. But reflection replaced defensiveness. Together, we created a simple, unobtrusive cue. When she noticed I was drifting into extended discussion during math time, she would quietly approach and ask, "What month is it?"

That single question became our shared signal. It reminded me of the instructional purpose of that moment. The time allocation was equally important in math and social justice. These conversations did not disappear – they were simply relocated to appropriate times. Her partnership sharpened my design. It strengthened my clarity.

This is the kind of professional collaboration that shapes teaching from the inside.

Teaching can be isolating work. Human partnership – colleagues who observe, question, and support – prevents drift. AI cannot replicate that relational exchange. It cannot offer perspective born of shared context. It cannot see what a teacher misses or gently signal realignment. It cannot strengthen craft through mutual trust.

Even in classrooms that are not technology-scarce, the principle remains: thoughtful design and human collaboration determine whether learning thrives. Tools matter. Access matters. But intention matters more.

AI can generate information. It cannot replace the humanity, partnership, and reflective dialogue that refine a teacher's practice.

Illustration in the Classroom: Elementary Level

Students are asked to design a model neighborhood.

One student begins by consulting AI. He asks what makes a good community. The response is organized and clear: schools, hospitals, parks, transportation, safety services. He transfers several of these features onto his planning sheet and begins constructing.

The model takes shape quickly. Roads connect. Buildings are labeled. The written explanation references "public services" and "accessibility." The project meets the checklist.

Then the teacher introduces a new condition. "Imagine you are eight years old and walking to school," she says. "Show me your route." The student traces a path with his finger. It crosses two major roads and passes directly beside a factory. "Where would you wait for the bus?" she asks. He pauses. The neighborhood was built from a list. It was not yet built from inhabiting.

The teacher shifts the task from placement to simulation. Students measure walking distances. They mark where traffic might accumulate. They test what happens if rain floods a low area. They revisit short readings about zoning and green space. Pieces move.

The hospital relocates away from industrial space. Trees are added between residential blocks and traffic corridors. A park shifts closer to apartments. The model becomes less symmetrical and more intentional.

AI provided structural categories. Design required students to think inside consequences. Teaching as delivery would have ended with labeled buildings. Teaching as design required students to inhabit the system they created and revise it under lived conditions.

Understanding did not deepen through more information. It deepened through reconfiguration.

Illustration in the Classroom: Middle School Level

Students begin a unit on ecosystems with a central question:
How does change in one part of a system affect the whole?
One student consults AI for an explanation of food webs. The response is accurate and complete. A labeled diagram accompanies the definitions. He copies key language into his notebook.

Then the teacher removes a predator from a modeled ecosystem and asks students to predict what will happen. "The prey population increases," he replies. "Show me," she says. The class moves into a simulation. Each student represents an element of the ecosystem and yarn connects energy pathways between them. When one predator steps out of the web, tension travels through the strands. Populations shift physically. Students must reposition to represent changes in available energy and space. The web slackens in some places and tightens in others.

Students begin adjusting their positions as energy pathways collapse. When resources thin, a student steps closer to another connection point. When overcrowding increases, movement becomes constrained. The static diagram transforms into kinetic reasoning.

After the simulation, students redraw their food web. This time arrows include feedback loops. Notes include variables: drought intensifies imbalance; human development reduces habitat; overpopulation stresses resource flow. The AI explanation described the system accurately. The simulation exposed its dynamics.

Teaching as delivery would have ended with definitions and diagrams. Teaching as design required constructing an environment where imbalance could be enacted, negotiated, and revised.

The concept moved from labeled relationships to experienced interdependence.

Illustration in the Classroom: High School Level

Students study originalism and the living Constitution.

One student consults AI for a summary. The explanation is balanced and precise. Definitions are cleanly articulated. Key cases are cited. Distinctions are organized clearly. The notes are solid.

Then the teacher introduces a contemporary case involving digital privacy and emerging technology. Students are assigned roles as Supreme Court justices. They receive briefs, precedent summaries, dissenting opinions, and constitutional clauses. They must now draft a majority opinion under time constraint. Definitions are no longer sufficient. Originalism demands attention to historical intent. The living Constitution invites adaptive interpretation. Both frameworks contain logic. Both generate tension when applied to technologies the framers never imagined.

In small groups, students deliberate. One argues that precedent anchors legitimacy. Another counters that rigid interpretation ignores evolving realities. They must anticipate dissent. They must articulate limits. They must justify trade-offs.

The student who relied on AI summaries begins weighing implications instead of reciting definitions. He considers how a ruling would affect future cases. He revises language to narrow scope. He adjusts reasoning to address unintended consequences. Interpretation becomes active judgment under constraint. The AI summary provided clarity. It did not conduct deliberation.

Teaching as delivery would have concluded with accurate notes on interpretive theory. Teaching as design required students to reason within competing principles, limited time, and public accountability. Understanding deepened not by adding more content, but by inhabiting consequence. Definitions describe frameworks. Deliberation reveals their limits.

When teaching is framed as delivery, success is measured by completion. AI excels in this model because it can transport information efficiently and fluently. If learning were merely the transfer of organized content, the task would already be optimized. Understanding deepens when students inhabit ideas, test assumptions, confront discrepancies, revise thinking, and defend reasoning in real time. Delivery ends when answers arrive. Design begins when students must use them.

Teaching as design shifts the central question from "What should students receive?" to "What must students practice?" It prioritizes sequencing, tension, reflection, and revision. It anticipates where misconceptions will surface and creates conditions where struggle becomes productive rather than punitive. It integrates emotional architecture with intellectual demand, ensuring students feel safe enough to risk being wrong while still being pressed to think deeply.

Participation does not happen by accident; it is engineered. When we design lessons that invite only the quickest hands or the most confident voices, we mistake visibility for engagement. Practical frameworks such as Persida and William Himmele's *Total Participation Techniques* remind us that intellectual inclusion can be structured deliberately – through think-pair-share variations, written quick responses, movement protocols, and low-stakes processing routines that require every student to generate thought (Himmele & Himmele, 2017). These structures do not replace relational teaching; they operationalize it. They ensure that participation is not performative but distributed, not dependent on personality but embedded in design.

Artificial intelligence can generate explanations, examples, and structures. It cannot orchestrate lived cognitive experience. It cannot sense when confusion signals growth. It cannot decide when to intervene and when to allow productive struggle.

In embracing design over delivery, teachers reclaim professional agency. They move from reacting to tools to intentionally structuring learning around them. AI becomes a material within the architecture – not the architect itself.

Learning does not occur because information was shipped successfully. It occurs because the environment was intentionally built for thinking.

Invitations for Preservice Teachers

Consider imagining teaching not as the delivery of information.

Instead imagine it as the design of learning experiences. When knowledge is instantly accessible, your value no longer lies in explaining content faster or more efficiently than a machine – it lies in crafting moments that invite curiosity, productive struggle, reflection, and growth. Designing learning requires intention. You decide what students will wrestle with, where they will pause, how they will collaborate, and when they will be asked to explain, defend, or revise in their thinking.

AI can generate answers fluently yet cannot cultivate understanding. That reality clarifies your role. You are not competing with AI's speed; you are designing the conditions under which insight becomes possible. You orchestrate emotional safety and intellectual challenge. You anticipate misconceptions and design for them. You build in opportunities for explanation, dialogue, and revision so that understanding cannot be outsourced.

As you enter the profession, trust that your creativity, judgment, and responsiveness are not optional enhancements – they are the work itself.

You are invited to consider: What would change in your planning if your first question was not "What will I cover?" but "What will students wrestle with?" How might your lessons look different if you designed for persistence rather than performance, for explanation rather than completion?

Teaching as design honors students as thinkers. It positions you not as a transmitter of information, but as an architect of understanding who constructs spaces where thinking can happen.

In an era of endless delivery, your capacity to design is what makes learning endure.

Invitations for Administrators

Teaching as design challenges systems which have historically organized around pacing guides, coverage, and compliance.

If teaching is reduced to delivery, then standardization appears efficient. But if learning is relational, interactive, and emergent then professional judgment becomes indispensable.

This reframing has leadership implications. When administrators protect time for planning, reflection, and instructional iteration, they are not indulging preference – they are safeguarding the conditions for meaningful learning. Design requires space for teachers to analyze student thinking, adjust tasks, collaborate, and refine instruction. In an AI era, leadership shifts from asking, "Was the content covered?" to asking, "Did students have the opportunity to think deeply?"

These commitments also unfold within real funding constraints. Budgets often favor visible purchases – curriculum packages, software licenses, assessment platforms – because they are concrete and easily justified. Time is harder to defend and easier to cut. Yet if teaching is design, the most strategic investment is not always a product; it is protected time and coherent support that allow teachers to think, revise, and grow.

You are invited to consider: What does your system signal about what matters most? Do your policies reward speed and compliance, or do they protect depth and professional judgment? Where might reallocating time – even more than money – strengthen the design capacity of your teachers?

Administrators who support teaching as design create coherence rather than fragmentation. By aligning policies, evaluation practices, and professional learning around depth rather than speed, they invest in teacher expertise and ensure that innovation strengthens learning rather than diluting it.

Leadership, in this sense, becomes architectural – designing systems that allow teachers to design learning.

Invitations for Homeschool Parents

Homeschooling naturally embodies teaching as design.

Without the constraints of institutional pacing, you already adapt learning to your student's interests, readiness, and questions. Such responsiveness is not informal – it is sophisticated instructional design.

When you choose to linger on a question, turn a curiosity into a project, or connect a math lesson to cooking, budgeting, or building something tangible, you are designing experiences that integrate thinking with life. AI may offer resources, generate explanations, or suggest activities, but it cannot determine what matters most. It cannot decide when your student needs challenge, reassurance, or rest.

Designing learning thoughtfully means asking not just, "What should my student know?" but "What kind of thinker is my student becoming?"

You are invited to consider: When do you intentionally slow down rather than move on? How do you decide when curiosity is worth extending, even if it disrupts your original plan? What signals tell you your student is ready to stretch – or needs space?

In embracing teaching as design, you help your student see education not as something delivered to them, but as something they actively participate in shaping. That shift – from passive receipt to active construction – is the foundation of lifelong learning.

Reflection Questions

Chapter 9: Teaching as Design, Not Delivery

1. If you stopped thinking of yourself as a content deliverer, what would change about how you design tomorrow's lesson?

2. What does your current lesson structure reward: speed, compliance, explanation, or reasoning?

3. When have you felt most like a designer of thinking rather than a manager of tasks?

4. Where in your lesson must students wrestle before they receive clarity?

5. If AI can generate the product, what part of the process are you intentionally designing to remain human?

Chapter 10: Questions Are the New Curriculum

If teaching is no longer delivery but instead is design, then we must then ask what exactly we are designing for. When information is abundant and fluency can be generated on demand, the center of gravity in classrooms shifts. We move from asking, "How do we transmit content?" to asking, "What kinds of thinking are we cultivating?"

That shift leads to a deeper realization: in an age where answers are everywhere, the real curriculum may no longer be answers at all. However, for most modern schooling, answers have been the currency of success.

Students earned grades by producing correct responses. Teachers demonstrated effectiveness by covering content; scope and sequence. Curriculum maps measured progress by how much information had been delivered and assessed. The underlying assumption was straightforward: if students could supply the right answers, learning had occurred.

Classrooms were often evaluated in terms of output rather than insight, speed rather than understanding. Students memorized, repeated, and performed, while teachers monitored compliance and correctness.

Artificial intelligence quietly disrupts that assumption.

When answers become instantly available, their value shifts. What once required memorization now requires discernment. What once signaled mastery now signals access. The scarcity that once gave answers their weight has largely disappeared.

An elementary student can retrieve historical facts in seconds. A middle schooler can generate a graph of a physics experiment with a click. A high schooler can draft a polished essay in minutes. These realities reveal something unsettling but clarifying – fluency is no longer reliable evidence of thinking. The presence of an answer does not prove the presence of understanding.

What remains – what cannot be automated so easily – is the ability to ask meaningful questions.

Questions shape thinking in ways answers never can. They reveal assumptions. They surface misconceptions. They invite curiosity and open space for multiple perspectives. A well-designed question does not conclude learning; it propels it forward.

In a landscape saturated with information, the quality of our questions becomes the true measure of intellectual depth.

I learned this before artificial intelligence entered my classroom.

It was my first formal evaluation as a teacher. I had prepared carefully. I created a detailed PowerPoint on exponents – clean slides, examples sequenced thoughtfully, practice problems embedded at just the right intervals, images I thought my students could relate to. I rehearsed transitions in my head. I felt ready.

When the bell rang, the internet did not. The screen remained blank.

There is a particular kind of quiet that settles over a classroom when technology fails. Students looked at me. My principal sat in the back with a clipboard. My slides – my structure – were gone.

So, I began to teach from memory.

I wrote on the board. I explained repeated multiplication. I modeled base and exponent notation. Students were attentive. They followed along, nodding as they solved problems at their desks. I felt the lesson steady beneath me. I had taught similar content before.

Until we reached exponent zero.

Without my slides to anchor me, I explained it quickly. I said that any number raised to the zero power equals zero.

The words left my mouth with confidence. They were wrong.

My principal interrupted the lesson gently but clearly. He stepped forward and said, "Actually, any nonzero number raised to the zero power equals one." He wrote it on the board beside mine.

There was no drama. No reprimand. Just correction.

The room was quiet again – this time differently.

Students erased their work. I nodded. I adjusted. We moved on.

But something inside me shifted.

I had delivered the lesson fluently. I had spoken with authority. Students had accepted my explanation without hesitation. No one had asked why exponent zero behaves differently. No one had pressed the logic. And I had not invited the question.

If a student had asked, "Why would any number to the zero power equal zero?" I might have paused long enough to reconsider. If I had asked myself, "What pattern are we noticing as exponents decrease?" I would have seen the error before speaking it.

Instead, I moved forward on performance.

That day, I learned something unsettling and clarifying: confidence is not the same as understanding. Fluency is not the same as precision. And authority does not guarantee accuracy.

The lesson students remembered was not exponent notation. It was watching their teacher be corrected – and continue.

In that moment, the most important shift was not mathematical. It was intellectual. We revised in real time. We acknowledged error publicly. We continued thinking.

Years later, when artificial intelligence began generating flawless explanations instantly, I thought back to that classroom. Answers can appear polished. They can sound certain. They can arrive without hesitation.

But certainty without inquiry is fragile.

What would have protected that moment was not better slides or stronger memory. It would have been a better question. Why does the pattern behave this way? What happens as exponents decrease?
What must be true for the pattern to remain consistent?

Answers conclude. Questions protect understanding.

Illustration in the Classroom: Elementary Level

Elementary

Students can already recite the life cycle of a butterfly's

life cycle. The sequence is secure. The diagrams are neat.

If asked to list the stages, every hand would rise. So the teacher does not ask for the list.

Instead, she asks, "Why is the chrysalis necessary?" The room pauses. A student offers, "So it can change."

"Change into what?" the teacher replies. "A butterfly." a student replies.

"What kind of change requires stillness?" the teacher presses.

The certainty of sequence gives way to speculation.

Students return to their observation journals. They compare early sketches to later ones. They note how the caterpillar's body reorganized. They recall the day the chrysalis appeared still – lifeless, even.

"What would happen if it skipped that stage?" the teacher asks.

The question unsettles the simplicity of the diagram.

Students begin hypothesizing. Without containment, wings might not form. Without protection, predators might interrupt development. Without stillness, transformation might remain incomplete.

The life cycle shifts from memorized order to functional necessity.

No new facts were added.

What changed was the question.

Teaching here does not replace knowledge. It stretches it. A sequence becomes a system. A stage becomes a condition for growth.

Overconfidence in recall dissolves when inquiry exposes structure.

The curriculum expands not by increasing content, but by increasing complexity.

Illustration in the Classroom: Middle School Level

Students are programming a robot to navigate a maze.

One team writes a straightforward sequence: Move forward. Turn right. Move forward. The logic seems sound. They press start. The robot moves confidently – then collides with the wall. "It should have worked," one student says.

The teacher does not explain the error. "What did your code assume?" she asks. They replay the sequence. The timing was fixed. The distance was guessed. The turn angle was estimated. The program assumed a world without variation.

The first success was not navigating the maze. It was noticing the assumption.

They revise. Conditional statements replace fixed timing. Sensor input replaces estimation. If the wall is detected within five centimeters, turn. If not, continue. They test again. The robot clears the first corner and misjudges the second. More revision.

"What did this version assume?" the teacher asks again.

Students begin tracing logic instead of blaming outcome. They identify where the code relied on ideal conditions. They refine parameters. They question thresholds. The maze remains unchanged. Their thinking does not. Debugging is not correction – it is exposure of hidden premises. Initial confidence yields to analysis.

The lesson is not about writing flawless code. It is about recognizing that every solution contains assumptions – and that design improves when those assumptions are tested.

Overconfidence dissolves under contact with constraint. Logic strengthens through iteration.

Illustration in the Classroom: High School Level

Students studying *Hamlet* can summarize the plot easily.

Themes are familiar: revenge, indecision, corruption, mortality. So, the teacher does not ask for a summary. Instead, she asks, "Does Hamlet's delay represent moral depth – or moral failure?" The room divides. One student argues that hesitation reflects conscience. Another counters that it reveals weakness. A third suggests the tension itself is the point.

The teacher complicates the frame. "If Hamlet acted immediately, would we admire him more?" "If he delays for moral reasons, why does he act impulsively elsewhere?" "Is indecision always weakness?"

The text reopens.

Students return to specific lines. They reconsider tone. They notice contradictions. Hamlet's reflection in Act III complicates his earlier resolve. His cruelty toward Ophelia disrupts a purely moral reading.

Interpretation, once confident, becomes layered.

The goal is not to choose a side. It is to hold tension.

Overconfidence in thematic labeling gives way to ambiguity. Claims must account for counterevidence. Certainty fractures into complexity.

The teacher does not resolve the debate. She sustains it.

Understanding matures when students learn that sophisticated interpretation does not eliminate tension – it articulates it.

Answers simplify. Questions destabilize.

Designing for interpretive tension reveals where reasoning must stretch.

Illustrations from the Classroom: University Level

Students present on the topic of AI in school systems.

Their slides are polished. Research is current. Definitions are precise. The arguments are coherent: efficiency, personalization, ethical safeguards. The presentations conclude. The teacher asks a different question. "What assumption about learning makes these benefits urgent?"

Silence. Students respond with familiar language – workload, differentiation, scalability. He presses further. "What if the urgency is not pedagogical, but systemic?" "What problem does AI appear to solve because of how we currently structure school?" "If we redefined the purpose of schooling, would the problem remain?"

The discussion shifts. Students reconsider their premises. Speed is assumed to equal effectiveness. Coverage is assumed to equal mastery. Efficiency is assumed to equal improvement. Those assumptions, once invisible, become visible. The conversation slows – not because students lack knowledge, but because their frameworks are being examined.

In subsequent weeks, the teacher assigns not conclusions, but destabilizing questions. Each student must identify a tension within their own argument. They must articulate what would challenge their position. "What is lost when personalization replaces shared struggle?" "Who benefits most from technological acceleration?" "What educational values are embedded in algorithmic design?" Overconfidence in policy recommendations gives way to philosophical inquiry.

Teaching is not just about adding more information. It is about interrogating the frame that organizes information. Complexity does not emerge by expanding content. It emerges by exposing assumptions. When students leave, the teacher does not expect them to remember every article. He expects them to remember how to question the premise.

When answers are abundant, the quality of questions determines the depth of learning. This is why questions – not content – are becoming the true curriculum.

AI can generate answers, but it cannot determine which questions are worth asking. It cannot recognize when a question exposes a flawed premise or invites deeper inquiry. It cannot sense the productive tension that signals intellectual growth.

Teachers can.

When educators prioritize questions, classrooms shift from performance to exploration. Students learn that thinking is not about arriving quickly at a conclusion, but about probing carefully and revising thoughtfully. Uncertainty becomes not a weakness to hide, but an invitation to investigate. Well-designed questions make thinking unavoidable. They require learners to analyze, synthesize, and reflect rather than merely record or reproduce information.

This shift reshapes everything – from lesson planning to assessment.

Instead of asking students to reproduce information, teachers ask them to interpret, compare, challenge, and connect. Instead of grading for correctness alone, they assess reasoning, reflection, and growth. Instead of measuring how much content students can recall, they examine how students think with that content.

Questions also humanize learning.

They create space for identity, experience, and perspective to enter academic discourse. A question such as, "Why might someone disagree with this interpretation?" cultivates empathy. "What's missing from this explanation?" fosters humility. "Who benefits from this outcome?" introduces ethical reasoning. These are not peripheral skills; they are central to civic and intellectual life.

AI cannot cultivate these capacities on its own. They emerge from relational, responsive, and reflective teaching. When students use AI to generate ideas, explore patterns, or test hypotheses, the real learning occurs in how they evaluate, critique, and refine those outputs – under the guidance of a thoughtful teacher.

When questions become central, AI shifts from shortcut to tool. Students may use it to generate multiple viewpoints, surface counterarguments, or identify patterns, but understanding develops through the discernment that follows. The tool may offer possibilities; judgment determines value.

The teacher's role, therefore, becomes even more intentional.

In an inquiry-centered classroom, teachers act as curators of thinking.

They design sequences of questions that deepen understanding over time. They listen closely to student responses and adjust in real time – pressing for clarification, challenging assumptions, extending reasoning. They model curiosity by asking alongside students, demonstrating that inquiry is communal rather than hierarchical. Classrooms become living laboratories of thought, where mistakes, revisions, and exploration are not interruptions but essential components of learning.

This approach prepares students not merely for exams, but for participation in a complex world. In workplaces, communities, and democratic societies, the most valuable individuals are rarely those who know the most facts. They are those who can ask the most incisive questions, frame problems thoughtfully, and navigate ambiguity with discernment.

Teaching through questions is not a retreat from rigor; it is its highest expression.

Across classrooms, centering inquiry over output, reflection over recitation, and dialogue over submission models the enduring human advantage: the ability to think critically, connect ideas meaningfully, and create understanding collaboratively.

In an age of intelligent machines, the future of learning belongs not to those who produce answers fastest, but to those who ask wisely.

Invitations for Preservice Teachers

In a world where answers are instant, the ability to ask meaningful questions matters. It becomes one of the most powerful acts of teaching.

Questioning not as a warm-up before "real" instruction begins, but as the core of learning itself changes learning. Strong questions do more than check understanding – they generate it. They surface assumptions, expose gaps in reasoning, and invite students to connect ideas across contexts.

When you prioritize questions, you shift the classroom dynamic. Students are no longer passive recipients of information; they become active participants in constructing meaning. You teach them how to analyze, how to challenge, how to sit with ambiguity long enough for insight to emerge. You model curiosity and intellectual humility, demonstrating that even the teacher is engaged in ongoing inquiry.

As you begin your career, remember that the questions you design communicate what you value. If your questions ask for recall, students will rehearse. If your questions demand reasoning, students will think. If your questions invite multiple perspectives, students will listen and revise.

You are invited to consider: What kinds of thinking do your questions require? Do they open space for exploration, or do they quietly signal that there is only one acceptable response? How might your planning change if you designed questions before designing activities?

AI can generate responses with remarkable speed, but it cannot cultivate wonder, discernment, or courage in the face of uncertainty.

That work – designing questions that stretch minds and shape habits of thought – belongs to you.

Invitations for Administrators

When questions become the curriculum, leadership must evolve alongside instruction.

Administrators are challenged to examine whether school systems truly reward inquiry or merely celebrate correct answers. Schedules, pacing guides, grading practices, and evaluation frameworks often signal that speed and measurable outcomes are prized above depth and exploration.

Change unfolds gradually. It begins with leadership clarity and is sustained through consistent, aligned decisions over time.

Leaders who value questioning intentionally create space for thinking. They protect time for discussion, seminar, and collaborative reasoning. They support assessments that require explanation and transfer rather than simple recall. They recognize that classrooms rich in inquiry may look less orderly on the surface – students debating, revising, reconsidering – but are intellectually vibrant underneath.

You are invited to consider: What do your systems quietly communicate about questioning? Do your policies create space for intellectual risk, or do they unintentionally reward quick answers and compliance? Where might small structural shifts – in scheduling, evaluation, or assessment – signal that depth matters more than speed?

In an AI era, leadership is not about ensuring every question has a rapid solution. It is about designing systems that honor the complexity of learning. When administrators model thoughtful questioning in meetings, policy decisions, and professional conversations, they reinforce a culture where inquiry is normative rather than disruptive. In doing so, they help schools move from measuring performance to cultivating discernment.

Invitations for Homeschool Parents

Homeschool Parents

Homeschooling offers a uniquely fertile ground for inquiry-driven learning.

Without rigid pacing constraints, you have the freedom to linger with a question that sparks curiosity or to follow an unexpected line of thought into deeper exploration. Some of the richest learning moments begin not with answers, but with wondering.

When you pause to ask "Why do you think that?" or "What might happen if...?" you teach your student that thinking is dynamic and relational. You model that uncertainty is not something to rush past, but something to examine. These conversations build habits of reflection and resilience that extend far beyond any single subject.

You are invited to consider: When was the last time you allowed a question to guide the day's learning? How do you decide when to follow curiosity rather than return to the plan? What patterns of thinking are you nurturing through the questions you ask at the dinner table, in the car, or during shared projects?

AI may provide information quickly and fluently, but it cannot sit beside a student in curiosity. It cannot notice the excitement in their voice or the confusion in their expression. It cannot tailor a follow-up question that connects a math concept to baking, budgeting, or building something in the backyard.

By centering questions in your home, you help your student develop confidence not only in what they know, but in how they think – an advantage that no tool can automate.

Reflection Questions

Chapter 10: Questions Are the New Curriculum

1. If answers are abundant, are your questions strong enough to matter?

2. What percentage of your classroom questions require defense, interpretation, or transfer – rather than recall?

3. What question once expanded your own thinking – and where do students encounter that same stretch in your classroom?

4. Whose questions drive the intellectual direction of your classroom – yours or your students?

5. If AI can answer your prompt instantly, what kind of question would require live thinking instead?

Chapter 11: Feedback, Not Grades

If questions reshape what we teach, they also reshape how we evaluate learning.

When inquiry becomes central, the purpose of assessment can no longer be to sort answers into right and wrong alone. It must help students refine thinking over time. In an era where answers are abundant and fluency is easily generated, the real measure of learning moves beyond the score and toward the conversation that follows.

Grades were never meant to carry so much weight.

They began as shorthand – a way to summarize progress, communicate expectations, and signal readiness to move forward. Over time, however, they became something else entirely: currency, identity marker, motivator, and verdict. For many students, grades stopped describing learning and began defining worth. As the National Research Council (2012) cautions, "When assessment becomes focused primarily on sorting and ranking students, it can undermine the development of motivation and productive dispositions toward learning" (p. 244). What was designed to inform growth gradually became a mechanism for comparison.

In today's classrooms, grades often function as both gateway and barrier. They shape access to opportunities, influence self-perception, and quietly inform students' beliefs about their intelligence and potential. What began as a reporting tool has evolved into a powerful psychological force.

Artificial intelligence makes this distortion harder to ignore.

When students can produce polished work with minimal effort, grades lose reliability as indicators of understanding. A high score may reflect access to tools rather than depth of reasoning. A low score may reflect fear, disengagement, or misunderstanding rather than limited ability. The distinction between what was submitted and what was actually learned becomes critical.

Grades tell us what was turned in. Feedback tells us what was learned. I learned this lesson while writing my dissertation.

I was the first in my direct family tree to pursue a Ph.D. That reality carried quiet weight. Years earlier, when I first began college, I had not even applied for my associate's degree, sure I would never return to university. My best friend still teases me about that decision – how certain I was that I did not belong in academic spaces long enough to claim even the credential I had earned. The memory is both humbling and instructive. Growth rarely announces itself at the beginning.

By the time I was writing my dissertation, the stakes felt different. The hours were long. I worked before sunrise and after the house was quiet. Each chapter felt like proof – not only of scholarship, but of persistence. I revised carefully. I cross-checked citations. I structured arguments with intention. Completion felt like accomplishment.

My Chair met with me regularly. She was direct in her critique and steady in her presence. She understood that dissertation writing is both intellectual and emotional labor – that identity and argument often evolve together.

One week, I arrived at our meeting with confidence, having completed a substantial section. It was organized. The literature was integrated. The pages felt substantial in my hands. I had already begun thinking ahead to the next section. In my mind, this portion was finished.

Her feedback was clear. The chapter did not work. It needed to be completely rewritten. There was no harshness in her tone. No dismissal of effort. Just precision.

For a moment, I felt the weight of the hours behind me. The drafting. The revising. What I had given up dedicating time to this writing. The quiet determination to do this well. My first instinct was to protect the work – to explain what I had intended, to preserve what I could.

But as she walked me through her thinking, I began to see it. I had summarized carefully. I had demonstrated command of the literature. I had written fluently. I had not yet made the argument. I had not connected with a throughline. The distinction was subtle and decisive. The writing was competent. It was not yet rigorous. It was organized. It was not yet sharp.

If that writing portion had received a grade, it might have been strong. The effort was visible. The structure was intact. Productivity could have been mistaken for progress. Instead, I was invited to begin again.

Starting over was not erasure; it was refinement. I returned to the literature differently. Not asking what it said, but what I was saying through it. I clarified my claim before drafting paragraphs. I cut entire sections that sounded intelligent but did not move the argument forward. The second version was more deliberate and with an unmistakable throughline.

The first draft had proven that I could write. The revision taught me how to think.

In that season of my learning, I found that feedback is not commentary on effort. It is an invitation to depth. It distinguishes between what is complete and what is clear. It refuses to confuse fluency with contribution.

Years earlier, I had underestimated my own trajectory, certain I would never need an associate's degree. Writing a dissertation required confronting that earlier certainty and outgrowing it. It required trusting feedback more than pride. Grades would have affirmed that I had finished. Feedback demanded that I become the kind of scholar who could write it well.

Feedback is where teaching lives: specific, timely, responsive, and relational. It names strengths, identifies areas for growth, and provides direction for revision. Feedback includes inquiry and responsiveness. Feedback is a dynamic, interactive interaction. Unlike grades, which imply finality, feedback assumes learning is ongoing. It positions mistakes as information rather than as judgment.

AI can assist with feedback at scale. It can generate suggestions, identify patterns of error, and provide immediate prompts. But it cannot replace the relational dimension that makes feedback transformative.

Students interpret feedback through trust. They receive it through the lens of relationship. Tone, timing, and presence shape how critique is heard. Some students lean into constructive challenge. Others withdraw until they feel understood.

Feedback is never neutral; it is relational communication. Its power lies not only in its content, but in its context. Recent research affirms this relational dimension for students: "affect-based trust ... predicts both their likelihood of seeking feedback and how they engage with it" (Mahbodi et al., 2025). When trust is present, critique feels like guidance. When trust is fragile, even helpful correction can sound like judgment. The words may be identical; the relationship determines whether they land as invitation or indictment.

I learned this lesson early – not in a classroom, but at home.

As a young parent, I began to understand something about feedback that no book or life lesson learned had fully prepared me for, and it did not land the same way twice. My firstborn son, Robert, was easily redirected as a toddler; a slightly raised voice was enough to capture his attention and interrupt his behavior. Yet I quickly learned that shift in tone could also affect him more deeply than I intended. I had to adjust my feedback delivery accordingly based on how it landed.

My second son, Richard, responded in nearly the opposite way. When my voice intensified, so did his – meeting volume for volume, intensity for intensity, escalating rather than pausing. Like a game of tug-o-war, Richard positioned himself to play the voice raising game. He was fully invested. I had to adjust my feedback delivery according to the impact it had for it to be effective.

My third child, Sunshine, reacted differently still; raised voices seemed to pass over her entirely, as if volume alone carried no persuasive weight. As if she were telling me raised voices did not deserve her attention. I had to adjust my feedback delivery to reach her in a way she could hear me.

My feedback in each experience was both a humbling learning experience and a time of reflection early in my parenting years.

Parenting is alike teaching in this way. Individualized adaptation is required for feedback to land effectively.

These differences were not inconveniences – they were instruction. Authority and intent alone was insufficient. What mattered was attunement: recognizing who was receiving the feedback, how they interpreted it, and what would invite reflection rather than resistance.

The same words, delivered in the same tone, produced entirely different outcomes depending on the listener. Over time, I learned to adjust not only my volume, but my pacing, posture, and language – sometimes kneeling to eye level, sometimes softening my tone, sometimes offering choice, sometimes allowing for natural consequences rather than any correction.

In doing so, I discovered that feedback is not simply correction; it is relationship in action. When delivered with awareness, it becomes formative rather than reactive. That lesson shaped not only my parenting, but my teaching: meaningful feedback does not assert control – it cultivates growth.

Feedback is effective only when it is responsive to the learner. It requires attention not only to what is said, but to how it is received. Learning depends less on uniform evaluation and more on thoughtful response.

When classrooms center feedback rather than grades, students shift from performance to practice. Mistakes become data rather than verdicts. Revision becomes expected rather than exceptional. Students begin to understand that learning is interactive, not evaluative.

This does not require eliminating grades entirely. It requires repositioning them.

In an AI-rich environment, feedback becomes a primary signal of learning. Teachers design tasks that require explanation, application, and reflection. They ask students to articulate reasoning, defend choices, and revise drafts. Feedback responds directly to those processes, guiding students toward deeper understanding rather than rewarding surface correctness.

This shift also restores integrity. When students know they will be asked to discuss, defend, and refine their thinking, outsourcing cognition becomes less appealing. The emphasis moves from producing a flawless product to engaging in a meaningful process.

In such classrooms, the most powerful assessment is not the letter at the top of the page – it is the conversation that follows.

Illustration in the Classroom: Elementary Level

A student struggles with spelling and sentence construction. Her narrative earns an 85 out of 100. The grade is solid. It tells her where she stands. It does not tell her where to go.

In the margins, the teacher writes lightly in pencil. Correct spellings are underlined – not as correction alone, but as markers of growth. Repetitive sentence beginnings are circled with a small prompt: "Another way?" Beside one vivid line about a dog chasing a ball, the teacher adds, "This helps me see it. Try one more detail like this." No lecture accompanies the marks.

During the next assignment, a similar note appears. "Strong verb." A week later: "Watch for repeated 'and then.'" Later still: "Dialogue adds movement." The comments are small. Consistent. Specific.

Over time, the student begins pausing before submitting her work. She rereads for repeated phrases. She experiments with dialogue. She adds a second descriptive sentence without being asked.

The gradebook shows modest variation – an 85, then 88, then 90. The more significant record is in the margins.

Early drafts show circled sentence starters. Later drafts show self-corrections before submission. In September, descriptive detail appears once per page. By December, it appears regularly.

There is no dramatic breakthrough. There is accumulation.

The teacher does not replace effort with polish. She does not overwrite the student's voice. She layers feedback gradually, allowing awareness to build.

Growth here is not an event. It is sediment.

The number marks performance. The margins mark movement.

Illustration in the Classroom: Middle School Level

A student submits a research paper on the Industrial Revolution. The structure is organized. Sources are cited. The grade is a 92 out of 100. The paper is strong.

In the margins, the teacher writes questions rather than corrections: "How does this evidence advance your thesis?" "What counterargument might exist here?" "Whose perspective is absent?" The student reads the comments and sets the paper aside. The grade feels sufficient.

A week later, during peer review, he encounters a similar question in a classmate's feedback: "Where is the analysis after this quote?" He rereads his own paragraph and notices that the quotation carries more weight than his explanation. He revises one section – adding two sentences of analysis.

On the next assignment, he includes brief commentary after each source. It is uneven but intentional. The teacher notes: "Stronger integration here." Another comment appears: "Push this further – why does this matter?"

The cycle repeats. Draft. Comment. Revision. New draft.

The grade remains high across assignments. The difference lies in density. Early paragraphs summarize events. Later paragraphs link cause and consequence more deliberately. Claims anticipate questions before they are asked.

In October, his thesis is declarative. By January, it includes qualification: "While industrialization increased economic output, it also intensified class inequities in ways reform movements struggled to address." The revision is not sudden. It is iterative. Each assignment carries residue from the previous one. Comments echo. Adjustments accumulate.

The teacher does not redesign the system. She sustains the cycle. Performance was already strong. Understanding becomes more textured through repetition.

Illustration in the Classroom: High School Level

A student submits an essay on memory and haunting in *Beloved*. The structure is controlled. The thesis argues that the ghost represents unresolved trauma. The analysis is competent.

In the margins, the teacher writes: "How does Morrison's structure shape our understanding of trauma?" "Does Beloved function only symbolically?" "What happens to identity when memory is fragmented?" The student revises a single paragraph, adding one quotation about narrative fragmentation. The thesis remains largely intact.

On the next essay, the teacher writes: "Notice how this narrator withholds information. What effect does that create?" The student experiments – tentatively – by analyzing point of view rather than plot.

Later in the semester, during discussion, a peer suggests that Beloved might represent collective memory rather than individual trauma. The student scribbles the idea in her notebook. When she returns to her essay portfolio at midterm, she adjusts her thesis slightly: "Beloved represents not only personal trauma but the inescapable persistence of collective history." The change is subtle.

Her paragraphs grow longer – not because of filler, but because she holds ambiguity longer before concluding. She integrates counter-readings. She questions her own claims in the final section. By spring, her essays do not resolve tension as quickly. They dwell. The grade trajectory is steady – A-, A, A. The intellectual texture changes more than the letter.

Interpretation deepens through revisiting, reframing, and re-reading. Each essay builds upon questions left open in the previous one. The teacher does not push for dramatic transformation. She sustains inquiry.

Over time, the student becomes more comfortable leaving space in an argument – acknowledging complexity without dissolving into vagueness.

Depth accumulates quietly.

Grades summarize output; feedback shapes understanding. Grades record performance at a moment in time. Feedback guides the next step.

When educators provide timely, specific, and relational feedback, students learn to analyze their own thinking, revise strategically, and deepen understanding. AI can assist with mechanics – flagging grammar errors or suggesting structural improvements – but it cannot replicate human judgment, relational trust, or interpretive coaching. It cannot sense defensiveness, encourage risk-taking, or tailor critique to a learner's readiness.

Recognizing this distinction reframes assessment. Success is no longer measured by correctness alone, but by the depth and quality of thinking that emerges through revision.

Parents intuitively understand this difference. A report card number tells them little about how their student approaches challenge, responds to critique, or perseveres through difficulty. Thoughtful feedback, however, opens dialogue. It explains not just what a student earned, but how they are growing.

Many systems rely on pre-populated report card comment choices for efficiency. While practical, and designed to save teacher time, such systems can limit the insight families receive about their student's intellectual journey.

Grades serve systems that require summary. Feedback serves learners growth.

In a world where polished answers are easy to generate, the real work of learning lies in interpretation, judgment, and revision. Feedback supports that work. Grades cannot.

Teaching has always been about helping learners grow. AI simply clarifies that growth cannot be reduced to a number without losing its meaning.

When feedback leads, learning follows.

Invitations for Preservice Teachers

Grades may summarize performance; feedback shapes learning.

As you enter the profession, it is tempting to see assessment as the final step in instruction – a score entered, a task completed, a unit closed. You are invited to think differently. Assessment is not the end of learning; it is the continuation of it. Every comment you write, every question you ask in conference, every moment you pause to say, "Tell me more about your thinking," extends learning beyond the product and into the process.

In an AI world, where polished paragraphs, correct calculations, and organized responses can be generated in seconds, feedback becomes the place where understanding is made visible. It is in dialogue – written or spoken – that you discover what a student actually grasps, where misconceptions linger, and how reasoning is evolving. When you design feedback that requires explanation, revision, and reflection, you signal that learning is iterative rather than transactional.

You communicate that growth matters more than appearance.

Your feedback also shapes identity. Students internalize the tone and focus of your responses. If feedback centers only on error, students may equate mistakes with inadequacy. If it highlights strategy, effort, and next steps, students begin to see themselves as capable thinkers in progress. In this way, your words do more than correct – they cultivate resilience and agency.

As you begin teaching, remember that feedback is one of your most powerful instructional tools. It guides growth, builds confidence, and reinforces that thinking – not perfection – is the goal.

In classrooms shaped by AI and accelerating change, your relational, intentional feedback becomes the anchor that keeps learning human.

Invitations for Administrators

Systems that rely heavily on grades as proxies for learning are incomplete. Grades offer efficiency and comparability. They satisfy reporting requirements and provide shorthand communication. Yet they can obscure the very elements that lead to meaningful understanding: struggle, revision, feedback, and growth over time.

In an AI era, this tension becomes more pronounced. When students can generate polished work with ease, grades alone become increasingly unreliable indicators of comprehension. Leadership must therefore ask deeper questions: What are we truly measuring? What do our assessment practices signal about what we value? Are we rewarding performance – or cultivating understanding?

Administrators who support feedback-rich classrooms take concrete steps to align systems with learning. They provide professional development focused on high-quality feedback. They examine evaluation frameworks to ensure teachers are recognized for designing revision cycles and facilitating reflective dialogue. They protect time for conferencing, collaborative scoring, and instructional reflection. They resist reducing assessment to data points detached from context.

You are invited to consider: Where do your grading structures unintentionally narrow learning? What conversations dominate staff meetings – point totals or evidence of growth? If a student improved dramatically but did not earn the highest mark, would your system recognize that progress? Consider shifting the conversation from "What grade did they earn?" to "What did they learn – and how do we know?"

When leaders elevate feedback as central to instruction, they create schools where assessment informs growth rather than merely labeling it. In doing so, they reinforce that learning is a process to be cultivated, not a product to be ranked.

Invitations for Homeschool Parents

Assessment often unfolds through conversation rather than marks on a page.

A raised eyebrow during a math explanation. A follow-up question after a history reading. A quiet, "Tell me how you figured that out." These moments are feedback. They shape learning in ways far more nuanced than a score ever could.

You may already practice this instinctively: provide feedback to help students understand not only what they did, but how they are growing. Without the pressure of institutional grading structures, you have the freedom to focus on revision, reflection, and progress over time. You can revisit a concept without penalty, extend a project because curiosity deepened, or slow down when confusion surfaces.

You are invited to consider: How does feedback sound in your home? Does it invite reflection or signal completion? When your student struggles, do your questions create space for growth or close the conversation too quickly?

AI may assist with checking answers, organizing information, or offering examples. But it cannot replace the attentive noticing you provide. It cannot observe frustration in a furrowed brow, celebrate effort in a hesitant attempt, or gently ask, "What might you try differently?" Your guidance transforms information into understanding by helping your student reflect on their thinking.

Through feedback, you teach something essential: learning is not a verdict. It is a journey shaped by iteration and growth.

When students experience assessment as dialogue rather than judgment, they develop confidence not just in what they know, but in their capacity to keep learning.

Reflection Questions

Chapter 11: Feedback, Not Grades

1. If grades disappeared, what would still guide your student's growth?

2. What does your current feedback communication about intelligence – fixed judgment or developing capacity?

3. When has your feedback changed the trajectory of a student's belief about themselves?

4. Does your feedback require revision – or simply signal right and wrong?

5. If AI can generate polished work, how does your feedback surface the student's actual thinking?

PART IV — THE TEACHER AS GUIDE, NOT GATEKEEPER

Chapter 12: Authority vs. Influence

As assessment shifts from a system of verdicts to a practice of dialogue, another shift quietly follows.

When feedback replaces mere scoring, the teacher's role changes. Authority no longer rests primarily in the power to assign a grade, but in the ability to guide growth. In classrooms where learning is iterative and relational, influence begins to matter more than position. It is from this shift – away from control as leverage and toward trust as foundation – that the next evolution in teaching emerges.

Evolution stems from history. For much of educational history, authority was assumed.

Teachers stood at the front of the room. Knowledge flowed in one direction. Compliance was expected, and respect was often defined as quiet obedience. Authority came from position, credentials, and control over information. This model functioned effectively when teachers were the primary gatekeepers of knowledge.

That context no longer exists.

There are still classrooms where hierarchical authority feels familiar, and for some educators, it remains the most comfortable instructional stance. Yet students have changed. The educational landscape has changed. Expectations within teacher preparation programs have also changed. Authoritarian delivery is rarely framed as best practice in contemporary training.

Artificial intelligence accelerates this shift because students can access explanations, examples, and alternative perspectives instantly. Authority can no longer rest on who knows the most. "AI makes expertise abundant and widely accessible, which shifts the role of the teacher from being the primary source of answers to being the guide who helps students make sense of them" (Mollick, 2023, p. 14).

Attempts to maintain control through restriction often backfire, breeding resistance rather than respect. Research on motivation confirms this dynamic: "Controlling contexts are associated with diminished intrinsic motivation and increased resistance" (Deci & Ryan, 2000, p. 236). When authority depends on gatekeeping information, it weakens in a world where information is ubiquitous. Authority grounded in discernment, however, grows stronger. It shifts from control to guidance, from enforcement to influence.

In this context, influence becomes central.

Influence is not granted by title; it is earned through trust, credibility, and relationship. Students listen because they believe the teacher has insight worth considering – not simply because they are required to comply.

Influence grows when teachers demonstrate intellectual honesty, curiosity, and ethical reasoning. It deepens when teachers admit uncertainty, invite dialogue, and guide students through complexity rather than shielding them from it.

This does not mean abandoning structure. Influence is not permissiveness. It is clarity paired with care.

Teachers who cultivate influence still define goals, establish boundaries, and hold students accountable. The difference lies in how those expectations are communicated. "Teachers become autonomy supportive when they identify, nurture, and develop students' inner motivational resources" (Reeve, 2012, p. 159). Shared purpose replaces fear. Dialogue replaces domination. Students comply not because they are controlled, but because they are engaged.

Authority commands compliance. Influence cultivates trust. And trust is where learning begins.

This distinction matters more than ever.

Teachers who rely solely on positional authority often experience AI as a threat – something that undermines their expertise. If authority is rooted in being the sole source of answers, then instant access destabilizes that foundation. In contrast, teachers who cultivate influence experience AI as a tool – something that can be integrated, questioned, and leveraged within a shared learning space. Their credibility rests not on exclusive knowledge, but on judgment, discernment, and the ability to help students navigate complexity.

Students are far more likely to deepen their respect for a teacher who says, "Let's figure this out together," than to lose it.

In classrooms grounded in influence, students are more willing to take intellectual risks. They ask questions without fear of embarrassment. They explain their reasoning, even when it is incomplete. They accept feedback because it comes from a relationship grounded in trust. Learning becomes collaborative rather than adversarial.

This model prepares students for life beyond school. Most adult environments do not operate on authority alone. Leadership in workplaces, communities, and families increasingly depends on persuasion, empathy, and shared vision. The ability to influence ethically – to listen, reason, and build consensus – is far more transferable than the ability to enforce compliance.

Teachers are uniquely positioned to model this form of leadership daily.

Artificial intelligence accelerates the shift by exposing the limits of positional authority.

When answers are pervasive, credibility comes from how one thinks, not simply from what one knows. Students pay attention to adults who help them interpret information, evaluate claims, and weigh consequences. They respect those who can hold structure and humility simultaneously.

In an age of abundant information, influence – not authority – becomes the enduring source of educational credibility.

Illustration in the Classroom: Elementary Level

Elementary

A student hesitates over a multi-step math problem.
The numbers feel crowded. The procedure feels rehearsed
but fragile. His pencil hovers.

The teacher does not restate the algorithm. Instead, she sits beside him and studies the page. "I'm not certain which path is simplest here," she says. "Let's look together." She circles the numbers. "What do you notice?" He shrugs. "What if we group these differently?" she suggests. "Let's test that idea."

She narrates her thinking aloud – not as instruction, but as exploration. "If I combine these first... that changes this part. Hmm. That's interesting."

When a miscalculation appears, she does not erase it quickly. She pauses.

"That didn't work the way I expected. What did it show us?"

The authority in her voice does not come from immediate precision. It comes from composure within uncertainty.

Peers offer alternate strategies. One groups by tens. Another subtracts first. The board fills with parallel reasoning paths. The teacher asks, "Which approach feels most efficient? Why?" The problem remains the same. The atmosphere changes from execution to examination.

Over time, the student begins verbalizing before solving. "I'm thinking if I do this first..." he says. When he makes an error, he looks at it longer before correcting it. He treats missteps as information rather than interruption.

The teacher's influence rests not in being the fastest solver in the room, but in being the most reflective one.

Authority can close uncertainty quickly.

Humility keeps it open long enough for thinking to mature.

Illustration in the Classroom: Middle School Level

Students investigate energy transfer, the experiment produces uneven results. Data points vary more than predicted. The expected pattern appears – but inconsistently. Students look toward the teacher. He studies the data table projected on the board.

"Our prediction was that increased surface area would accelerate heat loss," he says. "That seems mostly true – but not entirely." He pauses. "I'm reconsidering part of our hypothesis." The admission is deliberate.

He circles one anomalous data point. "If our thermometer calibration was slightly off, that might explain this value. Or we may not have controlled airflow carefully enough."

A student suggests that the metal container may have retained heat longer than expected. "That's possible," he replies. "Let's test that variable." Rather than restoring certainty, he models revision.

The class designs a second trial. This time, they adjust one condition. The results narrow slightly but do not align perfectly. He writes on the board: Hypothesis 1 → Revised Hypothesis → New Test.

Scientific reasoning becomes iterative and public.

Students begin proposing refinements more readily. They question initial assumptions without waiting for confirmation. The conversation centers on evidence, not correctness.

The teacher does not surrender authority. He demonstrates how it evolves.

Credibility grows when expertise includes recalibration.

The lesson is not that teachers do not know.

It is that knowledge itself is provisional – and responsible thinkers revise when warranted.

Illustration in the Classroom: High School Level

Students discuss *The Great Gatsby*. Several argue confidently that Gatsby represents the failure of the American Dream. The claim is supported with plot references and familiar quotations. The interpretation is coherent. The instructor listens. "I'm not sure that explanation is complete," she says.

The room stills – not because the claim is wrong, but because it is being stretched. "If Gatsby represents failure," she continues, "why does the narrative linger on his hope with such intensity? What would it mean if the novel critiques the structures around the dream rather than the dream itself?"

Students return to the text. One cites Nick's admiration. Another references Daisy's complicity. A third raises the tension between illusion and aspiration.

The instructor writes two columns on the board: Failure of the Dream. Persistence of Hope. Evidence accumulates under both headings.

When a student defends the original claim, she presses gently: "Where does your interpretation encounter resistance in the text?" The question is not corrective. It is complicating.

Discussion becomes layered. Gatsby's ambition appears both admirable and tragic. The American Dream becomes both possibility and distortion. The novel resists singular framing.

The instructor does not resolve the debate. She expands it. Her authority is visible not in delivering the definitive reading, but in refusing premature closure. Students begin drafting essays that anticipate counterarguments rather than eliminating them. Claims become more precise. Conclusions remain open to tension.

Humility here does not diminish rigor. It deepens it.

When authority shifts from position to influence, learning deepens.

Students respond differently when teachers guide reasoning, model inquiry, and invite collaboration rather than rely solely on directives. This distinction becomes especially visible when we consider motivation.

It is often tempting to secure immediate compliance through external incentives – offering candy for completed homework, extra screen time for quiet behavior, or game time for correct answers. These strategies can produce short-term results. Students comply. Tasks are finished. Order is maintained.

Yet as Smith, Fisher, and Frey (2015) caution in *Better Than Carrots or Sticks*, "Rewards and punishments may produce temporary compliance, but they rarely foster internal motivation or responsibility" (p. 18). The carrot may secure obedience in the moment, but it does not necessarily cultivate ownership. Systems built on external incentives can quiet behavior without transforming it. Lasting growth requires something deeper than control; it requires commitment.

What begins as a strategy for order gradually reshapes motivation itself. The consequences of this approach unfold quietly but predictably.

Over time, motivation shifts from purpose to payoff. The question becomes not "What am I learning?" but "What do I get?" Immediate compliance is achieved, yet influence is weakened. Students may meet expectations in the presence of rewards, but the deeper habits of self-regulation and intrinsic motivation remain underdeveloped.

AI accelerates the necessity of this shift. When access to answers is immediate, control over information no longer guarantees engagement. Neither increased surveillance nor stronger incentives sustain deep thinking. What remains powerful is the teacher's ability to shape how students think with that information – how they question it, test it, apply it, and critique it. Influence grows from trust, credibility, and modeling – not from restriction or reward. Teachers who understand this are empowered rather than threatened by technology.

They can integrate AI thoughtfully, design tasks that require judgment rather than recall, and cultivate dialogue that cannot be automated. In an AI-rich world, teaching is defined not by guarding knowledge, but by guiding discernment. The measure of expertise shifts from possessing answers to shaping the conditions under which wisdom can emerge.

Influence is quieter than authority, but it is far more durable.

Teaching in an age of intelligent tools is not about reasserting control; it is about redefining leadership. Teachers remain central – not as gatekeepers of information, but as guides in judgment, ethics, and meaning. Authority may command compliance. Influence cultivates trust. And trust is where learning begins.

I did not always understand the difference.

Early in my teaching career, I had a student who was the class comedian. He was quick – quicker than I was most days. His wit moved faster than my instructional pacing. He threaded humor through lessons, sometimes cleverly, sometimes disruptively. The class laughed easily with him. I could feel the room tilt in his direction.

At first, I appreciated his intelligence. Humor requires timing. It requires reading a room. It requires awareness. But gradually, something in me tightened. His interjections began interrupting explanations. His comments included teenage cultural references I did not recognize. I worried – privately – that I was missing something. That perhaps the humor carried a tone I could not fully decode. That it might border on disrespect or threat without my realizing it. Instead of asking, I reacted.

One afternoon, after a particularly well-timed interruption that derailed a key explanation, I corrected him publicly. My tone was sharper than I intended. I emphasized expectations. I reinforced boundaries. The room quieted immediately. Authority had worked. He did not interrupt again that day. But something shifted.

The next morning, he was different. The humor was gone, but so was the engagement. He stopped volunteering answers. He no longer leaned forward in discussion. When called on, he responded briefly and without expression. The class was easier to manage. It was also thinner.

Other students followed his lead. The laughter quieted, but so did the spontaneity. Fewer hands rose. Responses became cautious. The room complied, yet it no longer pulsed with energy. What I had interpreted as restored structure was, in truth, a narrowing of participation.

At the time, I told myself that structure had been restored. But gradually, I realized what had actually been altered. He had not withdrawn from learning. He had withdrawn from me.

In correcting him publicly out of frustration and uncertainty, I had protected order at the expense of relationship. My authority had secured compliance. It had not preserved trust.

What unsettled me most was not that he stopped joking. It was that he stopped thinking aloud. His quick wit had been connected to quick analysis. His humor, though inconvenient, had been a form of intellectual participation. When I silenced the surface behavior, I also dampened the deeper engagement.

Later, I reflected on what I had not done. I had not pulled him aside privately. I had not asked what he intended. I had not clarified expectations in a way that preserved dignity. I had acted quickly to reassert control, not slowly to cultivate influence.

The experience reshaped me.

In future moments like that, I learned to pause. To separate my insecurity from the student's behavior. To ask instead of assume. To redirect privately rather than perform correction publicly. To name expectations while protecting relationship.

Authority can quiet a room in seconds. Influence takes longer to build – and far longer to repair once strained.

In that classroom, I learned that leadership is not proven when students fall silent. It is proven when they continue to think, speak, and risk – even after being redirected. Trust is more fragile than order. And far more valuable.

Invitations for Preservice Teachers

Reconsider where your authority truly comes from.
In classrooms saturated with information, platforms, and AI-generated explanations, positional authority – standing at the front of the room, holding the credential, controlling the answer key – no longer carries the same weight. Students can fact-check in seconds. They can access alternative explanations instantly. If authority rests only on knowing more, it will feel fragile.

Influence, however, is durable. It grows through consistency, fairness, and visible integrity. Students follow teachers they trust – those who listen carefully, respond thoughtfully, admit mistakes, and model how to reason through complexity. Influence is built in small moments: how you respond when a student challenges an idea, how you handle uncertainty, how you treat the quiet learner who hesitates to speak.

In accelerated environments, your credibility does not depend on competing with technology. It depends on how you model judgment. When you say, "Let's examine this together," you demonstrate intellectual humility. When you ask students to defend their reasoning, you signal that thinking matters more than speed. When you acknowledge limits – your own and AI's – you teach discernment.

You are invited to consider: When students think of you, what do they trust you for? Your answers – or your reasoning? Your control – or your consistency? In moments of uncertainty, do you tighten authority or deepen dialogue?

Teaching is not about commanding attention; it is about earning it. And earning trust takes time. It is constructed through steady presence, ethical clarity, and relational consistency.

Invitations for Administrators

Leadership authority may come with a title, but influence must be cultivated. In schools navigating rapid technological change, mandates alone rarely create alignment. Policies can establish boundaries, but they cannot manufacture trust.

You are invited to reflect on how credibility is built – or eroded – through daily decisions. How are new initiatives introduced? Are teachers invited into dialogue before expectations are set? Is AI framed as a compliance issue or as a professional learning opportunity? In times of uncertainty, leaders who rely solely on authority often intensify resistance and anxiety. Those who lead through influence create shared purpose and professional confidence. Here, again, you are your teachers' most influential advocates.

In an AI era – where ambiguity is constant and tools evolve faster than policy – restraint, transparency, and listening become powerful leadership practices. Influence grows when administrators model curiosity instead of defensiveness, clarity instead of urgency, and respect instead of surveillance.

Teachers are far more likely to integrate emerging tools thoughtfully when they feel trusted rather than monitored.

You are invited to consider: When change arrives, does your first instinct lean toward tightening control or expanding dialogue? What signals do your decisions send about trust? In moments of tension, are you cultivating compliance – or confidence?

Influence at the administrative level sets the tone for the entire system. When leaders communicate openly, invite feedback, and acknowledge complexity, schools become environments where adaptation is collaborative rather than reactive.

Identify your leadership style. Notice where you default to authority. Then challenge yourself to cultivate the influence you want to see reflected in your teachers – and ultimately in your students.

Invitations for Homeschool Parents

Influence often matters more than rules. Students comply when required – but they internalize values when they observe them modeled. In a learning environment shaped by relationship, authority does not come from position alone; it comes from credibility and consistency.

Influence grows through everyday interactions: how you respond to questions, how you navigate disagreement, how you use – or limit – technology. When you pause to think aloud, when you admit uncertainty, when you revise your perspective in light of new information, you teach discernment.

In a world where AI can generate answers instantly, your example demonstrates something deeper: how to evaluate those answers, how to decide what matters, and how to act with integrity. Students learn not only from what you require, but from how you reason.

You are invited to consider: What patterns of judgment are you modeling in daily life? When your student watches you navigate uncertainty or disagreement, what do they learn about authority? Does your influence feel grounded in consistency and care – or in control alone?

When learning unfolds through conversation, shared problem-solving, and mutual respect, influence replaces coercion. And influence extends far beyond academics. It shapes how your student approaches authority, leadership, and responsibility in the broader world. In a culture saturated with information, your steady presence teaches them how to remain grounded in judgment, relationship, and trust.

Reflection Questions

Chapter 12: Authority vs. Influence

1. If students comply with you but do not trust you, have you actually led?

2. Where does your authority in the classroom come from – position, expertise, relationship, or credibility?

3. When have you seen a student follow your guidance because they trusted your judgment – not because they feared consequences?

4. Do your classroom structures cultivate student voice – or suppress it in the same of order?

5. In a world where information is instantly accessible, what makes your authority credible?

Chapter 13: Teaching Thinking, Not Tools

As classrooms shift from authority to influence, another subtle shift becomes necessary. If teachers are no longer gatekeepers of information but guides in judgment and meaning, then the central question changes. It is no longer, "What tool should we master next?" but rather, "What kind of thinking are we cultivating?" This distinction becomes especially urgent in an age defined by accelerating technological change.

With each new wave of technology, a predictable current pulls at schools: teach the tool. Districts rush toward innovation, eager not to be left behind. New platforms are layered onto old frameworks. Curricula bend to accommodate the latest software. Professional development fills with demonstrations, tutorials, and checklists.

Fluency becomes mistaken for formation. Mastery shrinks to mean competence within whatever system currently holds the spotlight. And quietly, almost unnoticed, the work of cultivating judgment gives way to training in technique.

I learned this lesson not as a classroom teacher, but as a principal.

At the time, I was leading a school serving incarcerated youth. The environment was structured, security-driven, and tightly regulated. Predictability mattered. Data mattered. Intervention planning mattered even more, and in this area we were limited. When I encountered a new digital platform that promised to track student behavior trends, generate intervention recommendations, and provide plug-and-play strategies aligned to evidence-based practices, I was immediately drawn to it.

The interface was clean. The dashboards were intuitive. Reports could be generated instantly. Patterns emerged visually in ways that felt actionable. The system translated behavior incidents into analytics and paired them with suggested responses. It appeared efficient, responsive, modern.

I adopted it enthusiastically.

Professional development resources were available. Staff learned to log incidents, categorize behaviors, and review automated recommendations. Within weeks, we had charts showing frequency patterns, heat maps of escalation times, and standardized intervention menus ready for implementation.

On paper, it looked like progress.

But gradually, something felt misaligned.

The platform treated behavior as data points. It sorted responses into tiers. It suggested structured interventions based on frequency and intensity. It assumed consistency of environment, predictability of routine, and a degree of emotional regulation not always available to students navigating incarceration.

Our students did not live in stable contexts. Their emotional landscapes were complex, shaped by trauma, systemic inequities, interrupted schooling, and confinement. A recommendation generated by an algorithm did not account for whether a student had just received difficult family news. It did not account for how power dynamics function inside correctional settings. It did not account for how trust is built differently in spaces defined by surveillance.

The tool functioned exactly as designed. It simply was not designed for our context.

What I had mistaken for innovation was, in reality, a misapplication. I had adopted efficiency without sufficiently interrogating fit. I had prioritized structure over discernment.

The realization did not come in a single dramatic moment. It surfaced slowly – in staff conversations, in student reactions, in interventions that felt procedurally correct but relationally hollow. Teachers began supplementing the recommendations with their own knowledge of students' histories. They quietly adjusted suggested strategies to account for trauma responses. The most effective interventions were not those generated automatically, but those shaped by human judgment authentically.

Eventually, I scaled back the platform's centrality. We retained useful data tracking features, but we stopped treating its recommendations as prescriptions. We re-centered relational understanding. Staff discussions shifted from "What does the system suggest?" to "What does this student need right now?"

That realization deepened when I began tracking something the platform could not measure.

At intake and again at exit, I administered the Children's Hope Scale, developed by Snyder and colleagues (1997), which defines hope as "a cognitive set that is based on a reciprocally derived sense of successful agency (goal-directed determination) and pathways (planning of ways to meet goals)" (p. 400). The measure captures not just optimism, but a student's belief in their capacity to generate routes toward goals and sustain motivation to pursue them.

Because of the school's location within a secure correctional facility, enrollment and incarceration were functionally inseparable. Time in the program meant time confined. I began examining hope scores alongside length of stay.

What I found was sobering.

Students' hope scores decreased after thirty days of enrollment. They declined further at sixty days. For those incarcerated ninety days or longer, the decrease was more pronounced still.

The behavioral platform categorized infractions and generated intervention menus. It translated complexity into charts and tiers. Yet it could not detect erosion of agency. It could not measure a shrinking belief in possibility. It could not account for how confinement reshapes a young person's sense of pathways forward.

The data compelled a question no software had posed: What does it mean to design interventions within a context that may itself be diminishing hope?

In that moment, the distinction between teaching tools and cultivating thinking became unavoidable. Systems can organize behavior; they cannot restore agency. They can produce efficiency; they cannot cultivate meaning. Judgment requires attending not only to what works procedurally, but to what sustains humanity.

The experience reshaped my leadership.

I began to see the limits more clearly. A tool can organize information, but it cannot interpret context. It can categorize behavior, but it cannot weigh lived experience. It can suggest interventions, but it cannot build trust.

I had adopted the platform believing it would strengthen our practice. Instead, it clarified something more essential: systems must serve thinking, not replace it.

The most powerful interventions in that school were never plug-and-play. They were relational, responsive, and deeply human. Technology offered structure. Leadership required discernment.

Artificial intelligence intensifies this same impulse – and exposes its limits even more quickly. Tools change too rapidly to anchor learning. What feels indispensable today may be obsolete tomorrow. Students trained narrowly in specific technologies risk becoming fluent in systems that no longer exist.

Thinking endures.

The purpose of education has never been to prepare students for a particular tool, but instead to prepare them for a world in constant change. That preparation depends not on technical fluency alone, but on judgment, reasoning, adaptability, and ethical awareness.

Even when schooling historically aligned with industrial demands – preparing workers for factories, offices, and standardized systems – the deeper purpose was adaptation. "Public schools have been asked to solve social and economic problems by adapting to new conditions while preserving familiar institutional patterns" (Tyack & Cuban, 1995, p. 84). Preparation has always meant equipping learners to navigate change, not merely comply with it. The tools evolve; the need for thoughtful adaptation remains.

AI makes this distinction unavoidable.

A student who knows how to operate a tool but cannot evaluate its output is not prepared. A student who can prompt a system but cannot detect bias, error, or omission has not developed independence. Teachers recognize this instinctively: the most meaningful learning moments emerge from wrestling with ambiguity, revising ideas, and transferring understanding across contexts.

Teaching thinking, therefore, means prioritizing transferable capacities: asking meaningful questions, evaluating evidence, recognizing bias, making decisions with incomplete information, explaining reasoning clearly, reflecting on mistakes, and adjusting strategies. AI can support practice in these areas – but it cannot internalize them on a student's behalf.

When teachers center thinking, tools become secondary rather than central. Students learn to approach technology with curiosity and caution. They learn to ask: "What does this tool help me see?" "What might it obscure?" "When should I rely on it – and when should I step back?" Technology becomes a resource, not a crutch.

This approach also restores equity. Tool-based instruction privileges access, bandwidth, and prior familiarity. Thinking-based instruction develops insight, perseverance, and perspective – capacities every student can cultivate with relational support.

When schools anchor learning in thinking rather than tools, they prepare students not just for the next update, but for a lifetime of change.

Illustration in the Classroom: Elementary School Level

Students are investigating a local environmental concern affecting a nearby river. The assignment is to research causes and propose recommendations. The teacher begins not with devices, but with a question. "What do we want to understand about this river—and why does it matter here?"

Students brainstorm possibilities: water quality, fish habitats, trash accumulation, flooding after heavy rain. One student sketches a map of the riverbank from memory. She remembers seeing a drainage pipe near a bend in the water. When laptops open, AI enters the process.

She asks for background information about river pollution in urban areas. Within seconds, a clear summary appears – common contaminants, runoff sources, ecological impact. The information is organized and helpful. It accelerates her entry into the topic. The summary provides vocabulary she did not yet have: stormwater runoff, sedimentation, watershed impact.

The teacher pauses beside her desk. "Which of these causes might apply to our river?" she asks. The student compares the generated information to her field notes. She circles phrases that match what she observed. She adds a margin note: drainage pipe near playground. The AI summary mentioned pollution broadly. It did not reference this specific feature.

She refines her research question. Instead of "How is pollution affecting rivers?" she writes, "How might stormwater from nearby construction affect our river's fish population?"

The tool accelerated orientation. The design of the lesson required localization. AI provided entry points and terminology efficiently. The student supplied context, specificity, and direction.

The final report integrates both – background research and firsthand observation. The assistant expanded access. The thinking shaped purpose.

Illustration in the Classroom: Middle School Level

Students analyze primary-source speeches from the Civil Rights Movement. The goal is not simply to summarize, but to interpret perspective and rhetoric.

One student generates an AI summary of a speech. The summary is coherent. It identifies major themes and historical context. It condenses emotionally charged language into neutral phrasing. It appears balanced and complete.

The teacher projects both the original speech and the AI summary side by side. "What did the summary preserve?" he asks. Students identify the main argument and key historical references. "What did it compress?" he continues. The room grows more attentive. They reread a paragraph from the speech. The original contains urgency, repetition, and metaphor. The AI version condenses it into two sentences.

Students begin marking differences. Where did emotional emphasis diminish? Which rhetorical devices disappeared? How does tone shift when language is neutralized? The AI summary is not inaccurate. It is selective.

The teacher invites students to annotate the summary – adding back rhetorical detail and contextual nuance. They note where audience reaction shaped the speech. They discuss how moral tension surfaces through repetition.

The tool functioned efficiently as an entry summary. The lesson emerges in the interrogation.

Students learn that summaries simplify by necessity. Interpretation requires restoration of complexity.

AI supports comprehension at scale. Discernment requires return to the original.

Illustration in the Classroom: High School Level

Students are modeling projectile motion in physics. Equations are known. Graphing tools produce elegant arcs within seconds. A student uses AI to calculate trajectory based on angle, velocity, and gravity. The predicted range is precise. The graph is symmetrical and clean. On screen, the model is persuasive.

Then the class conducts the physical trial. The projectile falls short. The arc is slightly uneven. The landing point varies across attempts. The teacher does not dismiss the model. She asks, "What conditions does this equation assume?" Students list them: no air resistance, consistent force, exact measurement. "And which of those conditions were present here?" Silence gives way to recalculation.

They measure launch angle more carefully. They estimate air resistance. They account for slight variations in applied force. The student overlays empirical data points onto the predicted curve. The discrepancy narrows – but does not disappear. The model remains valuable. It predicts within range. It clarifies relationships between variables. It is not reality.

The student writes in her lab conclusion: "The equation models ideal motion. Our results reflect environmental variables that the model simplifies."

AI accelerated computation and visualization. The experiment revealed constraint. Neither invalidates the other. Understanding emerges in comparison. The assistant provided speed and clarity. The classroom provided friction and calibration.

The lesson is not that models fail. It is that models require interpretation.

Teaching thinking is far more enduring than teaching tools. AI can generate content, organize data, or suggest solutions, but only students guided in reasoning, reflection, and evaluation develop transferable understanding. In classrooms that center thought over mechanics, tools become optional supports rather than crutches, and students learn to approach problems with discernment and curiosity.

When thinking is prioritized over tool mastery, students develop capacities that endure beyond any platform used.

Teaching thinking does not require abandoning technology. It requires reframing its role. Instead of asking students to use AI, teachers ask them to evaluate it. Instead of accepting outputs, they critique them. Instead of focusing on efficiency, they focus on understanding.

In doing so, teachers prepare students not just to use intelligent systems, but to live wisely alongside them.

The purpose of education, then, is not tool mastery – it is cultivating judgment, adaptability, and the ability to navigate complexity. AI highlights this truth: the future belongs not to those who master the fastest tool, but to those who think clearly when tools change.

In this moment, teachers are invited to reclaim the art of teaching, designing experiences that make thinking unavoidable, visible, and human.

That is the work of education. In this moment, it is simply impossible to ignore.

Invitations for Preservice Teachers

Look beyond the tools in front of you and focus on the thinking behind them. Platforms, programs, and technologies will continue to evolve – often faster than curriculum guides, standards documents, or district initiatives can respond. If you anchor your identity as a teacher to any one tool, you will constantly feel behind. But if you anchor it to thinking – to reasoning, reflection, and judgment – you will remain steady no matter how the landscape shifts.

Students do not need you to compete with AI. They need you to slow the moment down. To ask, "Why does this answer make sense?" "What assumptions are embedded here?" "What might we be missing?" Teaching thinking means designing lessons that require explanation, not just completion. It means asking students to defend conclusions, revise interpretations, and evaluate evidence. It means making their reasoning visible.

When intelligence is externalized, teaching thinking is an act of empowerment. When you prioritize metacognition – helping students understand how they think – you give them tools that no algorithm can replace. You prepare them to adapt, not just comply. To question, not just consume. To navigate ambiguity rather than fear it.

You are invited to consider: What habits of mind are your lessons quietly reinforcing? Are students practicing recall – or reflection? Completion – or discernment? If the technology disappeared tomorrow, what thinking would remain?

As you enter the profession, remember this: technologies will age. Software will update. Policies will shift. But the habits of mind you cultivate – curiosity, discernment, adaptability – will travel with your students long after the devices in their backpacks are obsolete.

Teaching thinking is not an accessory to instruction; it is the most future-ready work you can do.

Invitations for Administrators

You are invited to examine where attention, time, and resources are directed.

When systems invest heavily in tools without equal investment in thinking, learning becomes fragile – dependent on platforms rather than grounded in cognition. Schools can quickly become proficient in implementation while remaining shallow in understanding.

Administrators play a powerful signaling role. Agendas communicate priorities. Evaluation frameworks reinforce values. Budget decisions reflect beliefs about what matters. If training centers primarily on platform navigation, teachers may become efficient users of systems but hesitant designers of thought. If, instead, leaders protect time for lesson design, inquiry-based instruction, and reflective practice, they cultivate intellectual resilience across the school.

Leadership in an AI era requires discernment. Not every innovation requires immediate adoption. Not every tool deserves equal attention. The most forward-thinking leaders resist equating innovation with acquisition. They ask deeper questions: "How does this tool strengthen reasoning?" "Does it deepen inquiry?" "Does it support transferable understanding?"

You are invited to consider: Where does your professional development calendar point? What do your resource allocations quietly communicate about what you value? Are teachers being trained to operate tools – or empowered to design thinking?

When administrators prioritize cognition over convenience, they create schools that are not merely current – but enduring. In such systems, teachers are empowered to design learning that outlasts trends. Students graduate not just tool-ready, but thought-ready. That distinction determines whether schools chase change – or lead it thoughtfully.

Invitations for Homeschool Parents

Homeschooling offers a uniquely powerful environment for teaching thinking in real time.

You are not bound by pacing guides or bell schedules. You can pause when a question emerges. You can linger in confusion. You can follow curiosity where it leads.

Meaningful learning does not require mastering the newest platform. It requires reasoning, conversation, and reflection. Daily life presents countless opportunities to practice these habits: comparing prices at the grocery store, evaluating news stories, troubleshooting a broken appliance, debating ethical dilemmas in literature. Each moment becomes a laboratory for thinking.

AI may assist with information or practice. It can explain a concept, generate examples, or provide structure. But it cannot replace the thoughtful exchange that happens when you ask your student, "Why do you think that?" "What made you decide that?" "How else might we approach it?" Those conversations cultivate discernment. They build intellectual confidence. They teach students to trust their own reasoning while remaining open to revision.

You are invited to consider: Where in your daily routines can thinking be made visible? When your student reaches for an answer, do you pause long enough to explore the reasoning behind it? Are you modeling curiosity as often as you model certainty?

In nurturing these habits, you give your student something far more durable than technical fluency.

You help them become adaptable in a world that will continue to change. You model that learning is not about mastering tools – it is about mastering thought. And that gift will serve them long after specific technologies fade.

Reflection Questions

Chapter 13: Teaching Thinking, Not Tools

1. If the tools disappeared tomorrow, what would your students still know how to do?

2. Are students practicing decision-making or merely executing procedures in your classroom?

3. When have you witnessed a student think independently in a way that surprised you?

4. Where might your instruction unintentionally create dependence on prompts, models, or platforms?

5. Are you teaching students how to use tool – or how to decide when and why to use them?

Chapter 14: The Moral Compass of Education

As we move to examining what guides thinking, a deeper question emerges: toward what ends are we educating? If tools are not the center, and thinking is the aim, then we must ask what values shape that thinking. Intelligence without direction is not wisdom. Skill without reflection is not virtue. The conversation inevitably turns to ethics.

Education has always been a moral endeavor.

Every decision a teacher makes – what to include, what to omit, whose voices to center, how to respond to error – reflects values. These choices shape not only what students know, but how they come to understand fairness, responsibility, dignity, and power. Even silence communicates. Even structure carries belief.

Artificial intelligence does not change this truth. It sharpens it.

Any conversation about the moral compass of education must confront bias – both human and technological. Bias is not an exception to ethical teaching; it is one of its central tests.

Teachers, like all humans, bring lived experiences, assumptions, and cultural frames into the classroom. These influence how we interpret behavior, assess understanding, assign potential, and respond to students in moments of uncertainty. Ethical teaching does not demand the impossible standard of neutrality. It requires awareness, reflection, humility, and the courage to interrupt our own automatic judgments.

How we teach ethics inevitably reflects the lenses through which we see the world. I learned this unexpectedly in my forties, while preparing to meet an anesthesiologist before surgery. I had researched the procedure carefully and felt confident that I had weighed the pros and cons. What I had not considered was how my own unexamined assumptions might shape my emotional response; where my bias would appear.

The first time I thought about the anesthesiologist was when they entered the room, I felt my muscles relax and a quiet sense of reassurance settle in. The comfort did not come from knowing their credentials, understanding their expertise, or reviewing their professional success rates. It came from something more immediate and far less rational: I perceived them as male and

Asian. In that moment, I became aware of a bias I did not know I carried – that I subconsciously associated male, Asian medical professionals with competence. That recognition was humbling. It led me to wonder whether similar assumptions may have influenced how I interpreted, encouraged, or expected performance from my male, Asian students over the years.

Bias is not always loud or malicious; it can be subtle, embodied, and unnoticed until it surfaces.

This awareness matters even more in an age shaped by artificial intelligence. If individual human bias can quietly shape perception and expectation, technology trained on aggregated human decisions can amplify those patterns at scale. AI systems are built on historical data – data formed through decades of human judgments, inequities, exclusions, and systemic structures. As a result, AI does not erase bias; it can reproduce and legitimize it, often cloaked in the language of efficiency and objectivity.

When educators treat AI-generated outputs as neutral or authoritative, they risk outsourcing moral discernment to systems that cannot perceive context, culture, or lived experience.

The responsibility to examine bias, then, does not diminish with technological advancement – it deepens.

The danger is not simply that AI may contain bias. It is that its fluency can disguise it. Its confidence can obscure its limits. And its apparent neutrality can tempt us to suspend the reflective judgment that ethical teaching requires.

In an AI era, the teacher's moral compass does not become less necessary. It becomes indispensable despite its ongoing growth.

Once technology enters the classroom as a co-participant in learning, every decision about how it is used becomes an ethical one.

Teacher bias matters profoundly in this landscape. Decisions about when to trust a student, how to interpret AI use, whose work is questioned, and whose explanations are accepted are never value-free. Without reflection, implicit biases can shape which students are scrutinized more closely, disciplined more quickly, or assumed to lack understanding.

Ethical teaching demands that educators ask not only "Is this work authentic?" but also, "Why do I suspect this student more readily than another?" AI makes these questions unavoidable because it exposes how trust, credibility, and judgment are unevenly distributed.

Confronting bias, then, becomes a moral practice rather than a compliance task. It involves slowing down decisions, seeking additional evidence, listening carefully to student explanations, and remaining open to being wrong. It requires recognizing that ethical responsibility cannot be automated. No algorithm can weigh fairness, repair relational harm, or account for the long-term consequences of a disciplinary decision. Those responsibilities remain human – and they remain central to teaching.

In an AI-influenced educational environment, the moral compass of education must point not toward efficiency or certainty, but toward humility and care.

Teachers are called to notice where bias – human or technological – may distort judgment and to intervene thoughtfully. Doing so is not a distraction from learning; it is part of learning's ethical foundation. When educators model reflection, self-examination, and accountability, they teach students something far more enduring than content: how to navigate a complex world with integrity.

And students must learn to do the same.

Students need guidance to see and interrogate bias – both in themselves and in the systems they use. Teaching ethics in an AI age is not about instilling fear or issuing rigid rules. It is about cultivating discernment. It means helping students ask: "Who created this system?" "Who benefits from it, and who might be harmed?" "Whose perspectives are amplified, and whose are missing?" These are not abstract questions; they are civic ones, preparing students to navigate the moral terrain of society.

Teachers are uniquely positioned to guide this inquiry because ethical understanding emerges from dialogue, reflection, and exposure to multiple viewpoints. It deepens when students are invited to wrestle with ambiguity rather than avoid it. AI can simulate moral language, but it cannot feel responsibility, weigh consequences, or model accountability. Humans do.

Teachers who integrate ethical reasoning into daily instruction prepare students to engage responsibly with powerful tools. They demonstrate that efficiency is not the same as wisdom, and that innovation without reflection can deepen inequities rather than resolve them. This work requires humility.

Educators do not need to possess every answer; they must be willing to model uncertainty, to listen, and to acknowledge complexity. In classrooms where ethical reasoning is prioritized, students debate perspectives, confront contradictions, and practice navigating ambiguity with care.

AI supplies information. Teachers supply judgment.

Illustration in the Classroom: Elementary Level

The teacher gathers students on the carpet and writes a single word on the board: Fair. "What makes something fair?" he asks. Answers come quickly. "Everyone gets the same." "You take turns." "No cheating." The certainty is immediate.

Instead of affirming the first responses, the teacher introduces a design challenge. "We are going to write our classroom constitution," he explains. "A constitution protects rights and defines responsibilities. It tells us what we believe about fairness."

Students break into small groups to draft two statements: In our classroom, everyone has the right to... In our classroom, we agree to...

At one table, disagreement surfaces. One student insists fairness means equal turns in every discussion. Another argues that students who struggle should sometimes have more time or more support. The teacher does not resolve the debate. He draws a simple example on the board: one student wearing glasses to see the board, another not needing them. "Is that unfair?" he asks. Silence stretches.

Students begin revising their definitions. Fairness shifts from sameness to appropriateness. They consider safety, voice, and participation. They debate whether rights require responsibilities. They negotiate wording carefully – changing "everyone speaks the same amount" to "everyone has the opportunity to speak."

The constitution emerges gradually. It is handwritten, imperfect, and the result of compromise. No one leaves entirely satisfied. That is the lesson. Fairness is not obvious. It is negotiated.

The teacher did not deliver rules. He structured deliberation. Overconfidence in simple equality yielded to complexity in shared responsibility. The constitution on the wall reflects not compliance, but conversation.

Illustration in the Classroom: Middle School Level

Students examine historical documents. The assignment is not merely to summarize, but to evaluate competing claims about justice and strategy.

At first, interpretations are confident. One group argues that direct protest was the only morally defensible path. Another insists that incremental reform was more practical and therefore wiser.

The teacher introduces tension deliberately. “Was it ethical to break laws in pursuit of justice?” “Does moral urgency justify civil disobedience?” “What risks did each strategy carry?” Students must defend positions they did not initially choose.

The classroom reorganizes into structured debate. Each side presents a claim, anticipates a counterargument, and addresses potential unintended consequences. Observers track reasoning, noting where arguments oversimplify complexity. As discussion unfolds, students begin recognizing trade-offs. Direct protest created visibility but also backlash. Incremental reform reduced risk but sometimes preserved injustice.

The teacher intervenes not to settle the question, but to deepen it. “What assumptions about power are embedded in your argument?” “Who bears the cost of each strategy?” The debate becomes less about winning and more about weighing.

Students revise position statements before final submission, integrating opposing evidence. Some modify their original stance. Others strengthen it while acknowledging limits.

Ethical reasoning expands under tension. The lesson does not end with a single correct answer.

It ends with students recognizing that justice decisions require negotiation between principle and consequence.

Illustration in the Classroom: High School Level

Students analyze contemporary media narratives about a complex social issue. Initial responses are assured. Articles are categorized as biased or balanced. Arguments are labeled persuasive or flawed.

The teacher complicates the frame. "Assume each narrative is written by someone acting in good faith," she says. "How might each position be ethically defensible?" The room quiets.

Students are assigned to articulate the strongest version of an opposing viewpoint before defending their own. They must identify values underlying each claim—security, freedom, equity, stability. A student who initially dismissed one perspective as self-serving begins tracing the value logic beneath it. Another recognizes how a policy aimed at fairness may inadvertently produce new inequities.

The teacher maps competing values on the board: Security vs. Liberty. Equity vs. Efficiency. Stability vs. Reform. No value is erased. None is simplified.

Students draft analyses that acknowledge tension explicitly. Claims are qualified. Counterarguments are integrated rather than dismissed. Ambiguity is preserved instead of resolved.

The conversation shifts from "Which side is right?" to "Which values are in tension – and how do we decide among them?" Authority in the room does not determine the conclusion. It sustains the complexity.

By the end of the unit, students' writing reflects layered reasoning. Arguments include trade-offs, potential harms, and unintended consequences. Certainty has given way to calibrated judgment.

Ethical maturity does not eliminate conflict. It learns to reason within it.

Ethical education in the AI age is not about controlling technology; it is about cultivating discernment. Students naturally turn to AI for efficiency and clarity. Without guidance, they may confuse fluency with fairness or assume that polished language signals neutrality.

Teachers serve as moral anchors, modeling how to question authorship, recognize bias, weigh consequences, and reflect on responsibility. AI amplifies information – but it also amplifies existing inequities and assumptions. When educators weave ethical inquiry into daily instruction, they prepare students not merely to use technology competently, but to navigate it with integrity.

Ethical understanding cannot be outsourced.

Students require guidance to interrogate information, evaluate perspectives, and make decisions that account for fairness, responsibility, and power. Teachers cultivate this capacity through dialogue, reflection, and visible reasoning. They transform AI from a generator of unexamined answers into a catalyst for discernment. In a rapidly evolving technological landscape, the moral dimension of education becomes even more essential. Students depend on teachers to model judgment, interpret complexity, and demonstrate what responsible decision-making looks like in practice.

Education has always been a moral endeavor, and in that space, teachers remain irreplaceable.

Equity sits at the center of this conversation. Access to AI tools is uneven, and the consequences of misuse or limited access fall disproportionately on marginalized communities. As Darling-Hammond et al. (2020) observe, "The digital divide is no longer simply about access to devices, but about access to powerful learning opportunities supported by technology" (p. 21). Without intentional design, emerging technologies risk amplifying existing inequities rather than narrowing them.

Without intentional guidance, technology can widen gaps rather than close them. Ethical teaching requires naming this reality and responding thoughtfully. It demands attention to who has access, who benefits, who is scrutinized, and who may be harmed.

In an era of rapid technological change, the moral compass of education must point toward care, humility, and justice.

Students will encounter systems that shape opportunity, identity, and belonging. They need practice navigating those systems critically and compassionately. The goal of education is not to produce compliant users of technology, but thoughtful stewards of it. AI may shape the future – but values determine its direction.

This truth also calls educators to reflect on their own motivations. When individuals enter the profession with a desire to inspire understanding and help shape the adults of tomorrow, they align themselves with the deepest purposes of teaching. They also commit to hard work.

When teaching is treated as a fallback option, when it is "phoned in," considered an "easy" path, or something that can be done with minimal preparation and engagement, the work quickly unravels. In this context, "phoning it in" means approaching instruction with surface-level effort – doing the minimum rather than committing to the intellectual, relational, and ethical labor the profession demands.

Teaching has never been a profession where minimal effort suffices. In an AI age, that reality becomes unmistakable. Students no longer rely on teachers as the sole holders of information; they rely on teachers to notice them, challenge them, and bring meaning to the content.

AI may retrieve information faster. It may generate both accurate and inaccurate claims with equal confidence. Teachers are the ones who help students discern the difference. They teach students how to weigh evidence, question assumptions, and choose wisely.

The profession of teaching sits at the intersection of psychology, relational care, intellectual rigor, and creative expression. It requires wholehearted dedication and sustained reflection. Artificial intelligence does not diminish that demand – it intensifies it.

In this work, true impact comes not from efficiency or authority alone, but from deliberate effort and unwavering commitment to the learners in front of you.

Invitations for Preservice Teachers

Teaching has always been a moral act, whether it is named as such or not.

Instructional choices – what you emphasize, whose voices you elevate, how you respond to mistakes, how you distribute attention – communicate values. When tools can generate answers without conscience, context, or consequence, your role in shaping ethical judgment becomes even more essential. Students will encounter systems that sound confident, persuasive, and neutral. They need help to see what is missing, who is absent, and what is at stake.

You can teach not only content and skills, but discernment. Pause and ask: "Who benefits from this information?" "Whose perspective is centered?" "What assumptions are embedded here?" When you model those questions, you show students that intelligence without integrity is incomplete. You demonstrate that critical thinking includes moral reflection.

You are invited to consider: What values are embedded in the lessons you design? When students encounter certainty, do you teach them to look beneath it? When you select examples, texts, or technologies, whose stories are amplified – and whose remain unheard?

As you begin your career, remember that students are watching not only what you teach, but how you choose, question, and care. They notice whether you examine your own assumptions. They learn from how you admit uncertainty, repair missteps, and distribute fairness. Your moral presence – your humility, your courage to confront bias, your willingness to reflect – is part of the curriculum.

In an age of intelligent machines, your humanity is not a limitation. It is the compass.

Invitations for Administrators

Education is never neutral.

Policies, pacing guides, technology adoptions, and discipline structures all communicate what a school values.

In an AI-influenced landscape, administrative decisions carry profound ethical weight. How AI is introduced, who has access, how misuse is interpreted, and what outcomes are prioritized – all of these decisions shape the moral culture of schools.

Leadership in this era requires more than operational efficiency. It requires moral clarity. When systems reward output over reflection, surveillance over trust, or speed over equity, they quietly teach students what matters. Conversely, when leaders center equity, human dignity, and thoughtful judgment, they create schools where innovation strengthens rather than distorts learning.

You are invited to consider: What do your policies teach about dignity and fairness? When difficult decisions arise, what values guide your response? Do your systems communicate trust – or suspicion? Reflection – or reaction? In moments of technological uncertainty, are you designing for compliance, or for conscience?

You are invited to see yourself as ethical architects. Your commitment to fairness in discipline, transparency in policy, and equity in access becomes a lived model for the community.

When leaders align systems with ethical commitments – especially in times of rapid technological change – they provide a steady compass. In that steadiness, teachers feel supported rather than scrutinized, and students experience a school grounded not in fear of innovation, but in clarity of purpose.

Invitations for Homeschool Parents

Moral education is often inseparable from daily life.

It unfolds in conversations at the kitchen table, in how disagreements are resolved, in how sources are evaluated, and in how responsibility is practiced. Values are not an add-on to learning; they are its foundation. You know this. You may be homeschooling for these very reasons.

As new technologies enter your home, your guidance helps your student understand not just what tools can do, but what they should do – and when restraint matters. You are uniquely positioned to slow the moment down and ask reflective questions: "Is this fair?" "Who might this impact?" "What does integrity look like here?" In doing so, you teach discernment rather than dependence.

You are invited to consider: What ethical habits are being formed in the ordinary rhythms of your home? When technology offers convenience, do you also discuss consequence? When your student makes a decision, do you explore not only outcomes, but values?

By modeling empathy, responsibility, and thoughtful decision-making, you help your student develop an internal compass no algorithm can provide. In a world saturated with information and automation, wisdom will not come from speed or access. It will come from character.

Your steady presence – your willingness to discuss consequences, acknowledge complexity, and practice integrity – teaches something far more enduring than content. It teaches how to live well in a world of powerful tools.

Reflection Questions

Chapter 14: The Moral Compass of Education

1. What kind of human being is your classroom designed to produce?

2. When efficiency and equity conflict, which does your system quietly prioritize – and why?

3. What conviction about young people do you refuse to surrender – even when systems pressure you otherwise?

4. If your students adopt the habits your classroom rewards, what kind of citizens will they become?

5. Where might your daily practices contradict your stated educational values?

PART V — THE HUMAN SKILLS THAT MATTER MORE THAN EVER

Chapter 15: Attention in an Age of Distraction

If education is a moral endeavor – if teachers are called to cultivate discernment, integrity, and ethical judgment – then attention becomes one of its first responsibilities.

We cannot weigh consequences, interrogate bias, or practice reflection without the capacity to stay with an idea long enough to examine it. The work of moral education depends on sustained presence. And presence, in today's world, is increasingly fragile.

Attention is no longer assumed or guaranteed.

In today's classrooms, teachers compete with buzzing phones, open browser tabs, and the quiet pull of endless content. Outside of school, adults experience the same fragmentation: reading in half-pages, listening in partial sentences, living in interrupted moments. Artificial intelligence does not create this condition, but it accelerates it.

AI promises efficiency, instant answers, and frictionless interaction. It removes pauses that once invited reflection and shortens the distance between question and response, leaving little room for wondering, wrestling, or waiting.

But learning requires all three.

Attention is not simply focus; it is presence. It is the ability to remain with an idea long enough for understanding to deepen and for meaning to emerge. Without attention, thinking becomes shallow, reactive, and easily redirected.

Teachers have always worked against distraction, but the challenge now is not only external noise; it is internal fragmentation. Students are accustomed to constant stimulation. Silence can feel uncomfortable. Effort can feel unnecessary when alternatives are always available. AI intensifies this pattern by making cognitive shortcuts even more tempting.

Teaching attention in this context is not about enforcing stillness or banning devices. It is about helping students experience the value of sustained engagement. It is about designing learning that rewards patience, curiosity, and persistence rather than speed alone. When teachers slow the pace, invite reflection, and model deep listening, they communicate something powerful: this moment matters, and so does your thinking within it.

Attention also carries emotional weight. Students who feel unseen disengage; students who feel rushed withdraw. Presence communicates dignity. AI cannot offer that kind of attunement.

Teachers notice hesitation, confusion, and curiosity. They adjust tone, timing, and task in response to human cues. In doing so, they teach students not only how to focus, but why focus is worth protecting – and why sustained attention is essential for thoughtful, ethical, and meaningful learning.

Illustration in the Classroom: Elementary Level

The teacher introduces something small that does not look ambitious. Ten minutes. That is all. Ten minutes of independent reading at the start of each day – no devices, no audiobooks, no background music. Just books, laps, and quiet.

On the first morning, the room does not settle easily. A chair tips backward. A student whispers, "What page are you on?" Another flips pages rapidly without really reading. Eyes drift toward the clock after less than a minute. One child sighs dramatically. Another taps a pencil in irregular rhythm.

The teacher says nothing beyond the opening instruction: "Read." Three minutes in, she feels the pull to intervene. She recognizes the urge not as pedagogy, but as discomfort. Instead, she sits down with her own book. Five minutes feels long. When the timer rings at ten, several students exhale audibly. One says, "That felt like forever." The next day, they do it again. Nothing is different.

A week passes. The room still fidgets. The teacher notices that many students choose thin books with large print. A few abandon their selection after two pages. She does not scold. She quietly begins conferring during other parts of the day, helping students choose texts that stretch but do not overwhelm.

Two weeks in, the sighs soften. Not disappear – soften. The teacher introduces a small ritual: when reading begins, students place a colored bookmark at the page where their focus drifted the day before. The goal is not speed. It is returning. "Find where you left off," she says. "Stay a little longer today."

By mid Fall, the room still makes noise at first – a zipper, a shifting backpack – but it quiets more quickly. Students no longer look at the clock in the first minute. Some do, but not all. Ten minutes becomes twelve. The teacher does not announce the increase. She simply lets the timer ring later.

In late Fall, she notices something she cannot measure easily. When the bell rings unexpectedly one morning for a schedule change, several students look up startled – not relieved. One asks, "Can we finish the chapter later?"

The stamina is uneven. Some days feel regressive. Before long weekends, attention fractures again. After holidays, the first few days feel like starting over. The teacher does not interpret this as failure. She interprets it as muscle memory rebuilding. She continues.

By late Winter, the room holds silence more comfortably. Pages turn at irregular intervals rather than in restless flurries. A student who once tapped his pencil now rests his forehead lightly in his hand while reading. Another, who used to abandon books after a page, finishes one entirely and asks for another by the same author.

No celebration marks the change. No chart tracks minutes gained.

What has shifted is subtle but durable: students no longer experience quiet as emptiness. They experience it as space.

Artificial intelligence can summarize their books in seconds. It can generate comprehension questions instantly. It can even read aloud with perfect fluency. It cannot build the endurance required to sit with a story long enough for imagination to unfold. That endurance was not taught through a single inspiring lesson. It was built – minute by minute. Not through novelty. Through repetition. Through a teacher who chose not to compete with distraction, but to outlast it.

And in that outlasting, attention began to grow.

Illustration in the Classroom: Middle School Level

Students examine letters written by immigrants arriving at Ellis Island in the early 1900s. A student skims quickly, highlighting large sections of text without distinction. When discussion begins, her responses remain broad and general. The assignment is complete. The understanding is thin.

The teacher senses how easily he could praise completion and move on. Efficiency tempts him. Instead of correcting her interpretation directly, he changes the pace. He projects a short excerpt from one letter and reads it slowly, pausing mid-sentence. "Why does the author choose this word?" he asks. Silence stretches. "What does this tone suggest about how she felt leaving home?"

He underlines a single phrase – 'the streets feel louder than the ocean' – and returns to it, demonstrating how meaning unfolds through sustained attention. The room quiets – not because of enforcement, but because the pacing demands depth.

The teacher limits the task deliberately: three sentences, no more. Depth over volume. Students must explain why each sentence matters. They must justify their choices. Limitation becomes focus.

The student begins to notice subtleties she had skimmed past: a mix of hope and fear, coded references to discrimination, longing woven into ordinary description.

AI-generated summaries are available. They are clear and efficient, describing "immigrant challenges" in general terms. But the teacher reframes them as comparison tools rather than replacements. "How does the summary differ from what you noticed?" "What emotions were softened?" "What details disappeared?"

Artificial intelligence responds instantly because it does not experience effort. It cannot linger over metaphor. It cannot sense homesickness in a single line. It cannot model the discipline of rereading.

By the end of the unit, her contributions shift. She references precise phrases. She connects emotional tone to historical context. She questions omissions in summaries rather than accepting them at face value.

Middle School

The classroom culture reinforces that slowing down is not inefficiency – it is intellectual respect.

Attention, here, is not accidental. It is practiced.

Slowing down becomes a shared discipline rather than a teacher's demand.

Illustration in the Classroom: High School Level

A teacher makes a small decision that feels almost invisible.
She removes the slides. Not entirely – but mostly. Instead of a sequence of animated bullet points and embedded videos, she begins class with a primary source document projected in stillness. No music. No countdown timer. No immediate discussion prompt. Just text.

The first day, the silence feels awkward. Chairs shift. A student coughs. Another glances at a phone resting face-down but within reach. After thirty seconds, someone asks, "Are we supposed to be doing something?"

"Yes," she says gently. "Reading." They read for three minutes. It feels longer. When discussion begins, responses are shallow. Students summarize instead of analyze. Attention frays quickly; hands reach reflexively for distraction during pauses. The teacher notices the pull – the invisible gravity of constant stimulation – but she does not rush to compete with it.

The next day, the reading stretches to four minutes. A week later, five. Nothing dramatic happens. No sudden awakening. No cinematic breakthrough where students lean forward in collective revelation. Instead, something subtler unfolds.

The teacher stops answering immediately when silence falls. She lets the quiet settle. The silence unsettles her, too. She counts internally not only for them, but for herself. She resists the urge to rescue.

Students begin to realize that waiting will not produce entertainment. It will produce thinking. Some resist. One student mutters, "This is boring." The teacher nods. "Sometimes depth is quiet."

In mid Fall, she introduces handwritten annotations instead of digital highlighting. Phones are placed in a bin at the start of class – not as punishment, but as ritual. The first few weeks, students glance toward the bin as if something essential has been confiscated. By mid-semester, fewer eyes drift that way.

The class begins tracking how long they can sustain focused reading without interruption. Seven minutes becomes ten. Ten becomes twelve. A few students still struggle. That does not disappear. But more of them begin finishing passages without looking up. The change is not visible in test scores immediately. It appears instead in posture.

Shoulders remain squared toward the text a little longer. Students reference specific lines without prompting. During discussion, pauses stretch without panic. Someone says, "Wait – I need to think about that."

By late Winter, a visitor entering the room might not notice anything remarkable. Students sit. They read. They talk. But the teacher feels the difference.

Questions are more layered now. Students tolerate ambiguity longer before demanding resolution. When a primary source contradicts their assumptions, they stay with the discomfort instead of skimming past it.

Attention has not been captured. It has been trained.

There was no single moment when the room transformed. No applause-worthy strategy. Only repetition. Only consistency. Only the teacher's refusal to entertain away the silence.

Artificial intelligence produces without waiting. It can extract themes, generate arguments, predict likely interpretations. It cannot cultivate the stamina required to wrestle with complexity without acceleration.

That stamina was built here – minute by minute.

Not because the teacher dazzled. But because she stayed. And asked them to stay, too.

Attention is not enforced through control; it is cultivated through modeling, pacing, and relational trust.

Attention is not merely academic stamina. It is civic capacity. A mind trained only for acceleration cannot deliberate carefully, cannot hold competing truths in tension, cannot remain long enough to question its own assumptions. When teachers cultivate attention, they are not preserving nostalgia. They are preserving the conditions for thoughtful participation in a complex world.

AI can generate rapid answers, simulations, and summaries, but it cannot create the pause that invites depth. It cannot notice when a student is skimming past meaning or reward the discipline of careful thought. Teachers do that work. In doing so, they protect the space where learning shifts from speed to substance, from reaction to reflection.

Attention is neither automatic nor guaranteed; it must be intentionally developed. It grows through relational scaffolding, deliberate modeling, and structured reflection.

There was a moment early in my career when I thought I was managing a lesson efficiently.

Students were working independently. The room was quiet. I was moving between desks, answering questions quickly, keeping the pace brisk. We were covering material. Staying on schedule.

Then I noticed one student – usually engaged, usually thoughtful – sitting still, pencil in hand, unmoving.

I offered a rapid clarification and moved on. There were other hands raised. Other needs.

Later that day, she lingered after class.

"I didn't understand the first part," she said quietly. "But I didn't want to slow everyone down."

The words stayed with me.

I had been attentive to the task. Not to her.

My attention had been divided – between pacing, coverage, efficiency, the invisible pressure to keep momentum. I had mistaken motion for presence.

That moment unsettled me more than any overt disruption. Distraction is not always noisy. Sometimes it hides inside productivity.

In the years since, I have learned to watch for subtle withdrawal – the student who stops asking questions, the eyes that glaze just slightly, the silence that signals confusion rather than concentration. Slowing down feels risky in a culture that prizes coverage. But depth requires it.

Artificial intelligence accelerates many things. It cannot replace the attuned pause – the moment when a teacher senses that understanding is thinning and chooses to remain.

Attention is not simply about resisting devices. It is about choosing to stay.

And staying, in a world built for speed, is an act of care.

That act of care becomes even more essential in an environment shaped by artificial intelligence.

AI offers efficiency and immediacy, but it cannot inhabit the classroom, read hesitation, or sense when engagement is thinning. When teachers slow the pace, attend to student cues, and design tasks that require sustained thinking, understanding deepens. Attention becomes both a cognitive skill and a moral practice – teaching students that presence, focus, and thoughtful reflection are worth protecting.

In a world saturated with distraction, the classroom remains one of the few spaces where attention is practiced deliberately. Teaching students to attend – to ideas, to people, and to themselves – is preparation for a meaningful life. It strengthens reasoning, deepens relationships, and fosters agency.

AI may optimize speed. Learning thrives on depth.

And the classroom, guided by human presence, remains one of the last places where depth is still intentionally cultivated. That alone makes teaching indispensable.

Invitations for Preservice Teachers

Attention is not something students either possess or lack;
it is cultivated.

It is shaped by environment, relevance, emotional safety, and the relationships that frame learning. You are invited to move beyond the belief that focus can be demanded through rules or restored through reprimands. In an age of constant distraction – notifications, open tabs, algorithm-driven feeds – your role is not to compete with stimulation, but to design learning that is worthy of sustained engagement.

When lessons are purposeful, cognitively challenging, and emotionally safe, attention follows. Students attend when they feel seen. They attend when tasks require thinking rather than passive completion. They attend when they understand why something matters. Designing such environments requires intentional pacing, clear structures, and meaningful questions that slow thinking down rather than rush it along.

You are invited to consider: What in your classroom invites attention rather than demands it? Do your structures cultivate curiosity, or merely compliance? When distraction appears, do you tighten control – or deepen purpose?

AI may offer speed and instant gratification, but you offer something different: presence. You model what it looks like to linger with an idea, to wrestle with uncertainty, to revise a thought before speaking.

When you narrate your thinking, pause for reflection, and protect silence as productive space, you teach students that depth is not accidental – it is chosen. In doing so, you cultivate not just academic focus, but the lifelong habit of sustained attention that meaningful learning requires.

Invitations for Administrators

Attention is shaped not only in classrooms, but by systems.

Examine how school structures either protect or fragment cognitive focus. Schedules packed with constant transitions, overlapping initiatives, excessive compliance demands, and a culture of urgency communicate that speed is valued over depth. In an AI era – where distraction is ambient and immediacy is normalized – leadership must intentionally defend sustained engagement.

Protecting attention means more than limiting devices. It means creating coherence. It means aligning initiatives rather than layering them. It means ensuring that teachers have uninterrupted planning time, that students have extended blocks for deep work, and that reflection is built into the rhythm of the school day. When leaders prioritize realistic pacing and instructional clarity, they reduce cognitive overload for both teachers and students.

You are invited to consider: What in your system fractures attention? Where has urgency replaced depth? Do your schedules, expectations, and initiatives cultivate sustained thinking – or constant reaction?

Administrators who value depth over immediacy send a powerful message: attention is not a behavioral problem to manage; it is a resource to steward. In doing so, they create conditions where thinking can unfold fully rather than fragment under pressure.

In a landscape of accelerating tools and information, this protective leadership becomes essential to preserving the integrity of learning.

Invitations for Homeschool Parents

Homeschooling offers rare gifts. One is the ability to align learning with natural rhythms of attention. Focus grows when learning is meaningful, relational, and thoughtfully paced. Without rigid bells or crowded transitions, you can recognize when your student is ready to immerse themselves deeply in a subject – and when it is time to shift toward movement, dialogue, or restorative pause.

AI may capture attention quickly, but it does not teach how to sustain it. Sustained attention develops through shared reading, extended projects, reflective dialogue, and quiet persistence. When you linger over a question, invite your student to explain their thinking, or revisit an idea over several days, you are strengthening their cognitive endurance.

You are invited to consider: What rhythms of attention shape your days? When does your student seem most deeply engaged – and what conditions made that possible? Do your routines protect depth, or drift toward distraction?

Through your presence and intentional routines, you model that attention is not about speed or productivity; it is about care. You teach your student that some ideas deserve time, that understanding requires patience, and that focus is a skill worth practicing.

In nurturing these habits, you prepare them not only for academic success, but for a life marked by depth, creativity, and thoughtful engagement in a world that rarely slows down.

Reflection Questions

Chapter 15: Attention in an Age of Distraction

1. What in your classroom design competes with the very attention you are trying to cultivate?

2. How long are students asked to sustain one line of thinking before shifting tasks?

3. When have you experienced sustained attention as transformative – and where do students experience that same immersion in your classroom?

4. Are your technology routines training divided attention or disciplined focus?

5. What beliefs about their own ability to concentrate are students forming in your classroom?

Chapter 16: Belonging, Identity, and Voice

Sustained attention creates the conditions for thinking to deepen. Yet even the most carefully designed learning environment cannot flourish if students feel invisible within it. Focus depends not only on cognitive engagement, but on emotional safety. Before students lean into complexity, before they risk asking questions or revising ideas, they make a quieter assessment: Is this a place where I am seen? Do I belong here?

There was a time in my teaching when I believed I was cultivating strong academic discussion.

Students were participating. Hands rose. Responses were articulate. The room felt engaged. I mistook fluency for belonging.

One student rarely spoke. He completed assignments carefully, submitted thoughtful written work, and maintained steady eye contact when addressed directly. But during whole-group dialogue, he remained quiet. I interpreted his silence as preference. Perhaps he was introverted. Perhaps he processed internally. Perhaps he understood something about the room that I had overlooked. I did not press.

One afternoon, during a small group activity, I overheard him explaining an idea to a peer with precision and depth I had not yet heard in the full class setting. His reasoning was layered. His insight nuanced. His voice confident.

When I asked later why he did not share those ideas publicly, he shrugged.

"I don't always think people want to hear it like that," he said. "It sounds different."

Different how? He gestured vaguely. "Just different."

The room I thought was inclusive suddenly felt less certain. Whose language patterns had I affirmed? Whose examples had I amplified? Whose phrasing did I paraphrase more enthusiastically? I began replaying past discussions. I noticed how often I had reworded certain students' comments into more "academic" language before validating them. I noticed which dialects I left untouched and which I subtly reshaped.

I had not silenced him. But I had not explicitly legitimized him either.

Belonging is not established by absence of harm. It is established by presence of affirmation.

From that year forward, I changed small things that altered the climate in my classroom. I named multiple forms of expression as academically valid. I asked students to explain ideas in the language that felt natural before translating into formal academic phrasing. I modeled how to hold space for dialect and discipline simultaneously. I listened for the ideas beneath unfamiliar phrasing rather than rushing to refine it.

Gradually, his participation shifted. Not dramatically. Not all at once. But with increasing steadiness. His contributions entered the room in his own cadence. And when peers began referencing his insights later in discussion, something settled.

Belonging had moved from theory to architecture.

It is possible to build a classroom that values achievement yet quietly centers only certain forms of voice. It is possible to maintain high standards while unintentionally narrowing legitimacy. I learned that cultivating belonging requires more than inviting participation; it requires examining whose participation feels safe without translation.

Before students risk complexity, they decide whether their thinking can exist without erasure.

Belonging answers that question.

If attention creates the conditions for depth, belonging creates the conditions for voice.

And voice cannot grow where legitimacy is uncertain.

It is here that the next layer of teaching emerges – not simply protecting attention, but cultivating belonging.

The work is not about comfort alone. It is about legitimacy. It is the sense that one's presence is welcomed, one's perspective matters, and one's voice will not be dismissed or diminished. When students feel connected to their school, they are "more likely to engage in learning and less likely to engage in risky or disruptive behaviors" (Zins et al., 2004, p. 3). Participation grows from that connection. When belonging is absent, students often protect themselves – through silence, defiance, or disengagement.

Academic risk requires relational safety. Artificial intelligence complicates this landscape.

AI can generate language that sounds confident and polished, often mirroring dominant norms of expression. For students whose voices have historically been marginalized – because of culture, language, disability, or identity – this can feel both empowering and erasing. AI may provide access to academic language and conventions, yet it can also flatten nuance, smoothing out dialect, storytelling traditions, and culturally rooted forms of expression.

When fluency becomes the standard, authenticity can quietly disappear.

Belonging requires more than fluency. It requires recognition.

Teachers play a central role in shaping whether classrooms affirm identity or subtly pressure conformity.

This work is not performative; it is relational. It shows up in whose ideas are invited into discussion, how contributions are reframed or expanded, whether multilingual expression is valued as asset rather than deficit, and whether students see their lived experiences reflected in curriculum and conversation. It is present in tone, timing, and the small acknowledgments that communicate, You matter here.

AI cannot do this work. It does not see students. It does not know their histories, hopes, or hesitations. It cannot sense when a student's silence signals uncertainty rather than indifference. It cannot notice the courage required for a student to share a personal perspective. It cannot create spaces where vulnerability feels safe.

Voice develops in relationships. It grows where dignity is affirmed, where presence is recognized, and where belonging is practiced rather than proclaimed.

Illustration in the Classroom: Elementary Level

Students sit in a circle and share responses to a simple prompt: Describe a place that feels important to you.

Hands rise quickly. Stories spill easily – a grandmother's kitchen, a soccer field at dusk, a favorite lake. One student remains quiet. When her turn comes, she shrugs. "I don't know." The teacher moves on without pressure.

Later that afternoon, she replays the moment. The student had written a full paragraph during independent work the day before. The language was vivid. The thinking was precise. Silence, she realizes, is not absence. She begins paying closer attention.

Over the next week, she notices a pattern. The student writes readily but rarely volunteers to speak. When others share stories that feel familiar to the teacher – vacations, large family gatherings, summer camps – the teacher responds enthusiastically. When stories reference places she has never known – apartment complexes, shared bedrooms, neighborhoods described with guarded language – she tends to move more quickly.

She recognition is uncomfortable. Belonging, she realizes, is not simply about inviting students to speak. It is about which experiences linger in the room. The next morning, she changes the structure slightly. Students write first. Then they share in pairs. Only afterward does the circle open for volunteers.

When the quiet student speaks to her partner, the teacher kneels nearby, not hovering but listening. The story is about a small balcony outside her apartment – the only place she feels alone. She describes the sound of traffic below and how she watches the sky change colors at night.

During whole-group sharing, the teacher does not summarize the student's story into something more familiar. She asks a simple question: "What makes that space yours?" The student answers softly, but without apology.

Over the next few weeks, participation remains uneven. Some days she speaks. Some days she does not. But when she writes, she writes as someone whose experiences are worth describing. When she listens, she listens as someone whose interpretation matters.

The teacher becomes more deliberate about what she amplifies. She lingers over details that differ from her own experience. She resists translating them into something more comfortable. She notices which students are interrupted and which are not. She adjusts seating subtly, redistributing proximity and attention.

Belonging does not arrive as applause. It appears in small shifts: a raised hand without visible bracing, a story told without preemptive laughter, a peer asking a follow-up question instead of changing the subject.

Artificial intelligence can generate stories in any voice. It can simulate dialect, setting, and tone. It cannot feel what it is like to risk offering one's own.

Belonging is not a performance of inclusion. It is the lived experience of safety under observation. When students do not feel they belong, they do not simply withdraw. They withhold perspective. Entire ways of seeing remain unspoken. In this classroom, the teacher is not manufacturing confidence. She is adjusting the conditions under which identity is allowed to surface.

The work is ongoing. Belonging is not declared once. It is signaled repeatedly. And in those repeated signals, students begin to understand not only that they are welcome in the room – but that their way of seeing has a place within it.

Illustration in the Classroom: Middle School Level

Students prepare for a structured debate on a contemporary policy issue. They have read multiple perspectives, annotated arguments, and gathered evidence. The structure is clear. The expectations are rigorous. When discussion begins, a familiar pattern emerges.

A small group of students enters quickly – confident, fluent, comfortable interrupting. Their voices overlap easily. They challenge one another with visible ease. The tempo of the room accelerates around them. Others sit upright but silent.

One student in particular writes constantly in the margins of her notes. Her annotations are layered, careful, nuanced. But when the floor opens, she lowers her eyes. When invited in, she offers short responses – accurate, but restrained. From a distance, the debate appears vibrant. The teacher feels a flicker of satisfaction. The energy is high. The arguments are sharp.

Later, reviewing the discussion transcript, he notices something he had missed in the moment. The same four students spoke nearly seventy percent of the time. The quiet student's written analysis – collected afterward – is among the most complex in the room.

He reconsiders what he had equated with engagement. The next day, he alters the structure. Students write a full claim before speaking. Then they must identify which classmate's idea they are building on. Interruption is replaced with sequence. Speed is replaced with citation. At first, the dominant voices feel constrained. The quieter students appear uncertain about entering the space, even with structure. The teacher resists rescuing the silence.

When the quiet student speaks, she references a line others overlooked. She does not raise her volume. She does not perform conviction. She simply traces the logic of the argument and names its weakness.

The room pauses – not out of politeness, but because the point requires reckoning.

Over the following weeks, patterns shift slowly. The most confident students learn to listen for entry points rather than seize them. The quieter students begin to speak earlier in the discussion, not waiting until conclusions have already hardened.

The change is uneven. Some days regress. On days when energy runs high, familiar voices still dominate. But the teacher notices something subtle: the quiet student's posture changes. She enters with her pen already uncapped, as if anticipating participation rather than bracing for it.

Belonging here is not volume. It is intellectual legitimacy.

Artificial intelligence can generate arguments from multiple sides of an issue. It can simulate debate structure and produce counterclaims instantly. It cannot feel the social risk of entering a contested space with one's own reasoning.

When students do not feel they belong in academic discourse, they do not merely fall silent. They begin to believe their analysis is peripheral – that thinking belongs to others.

The teacher's work is not to make every voice loud. It is to ensure that no voice becomes invisible. Belonging, he learns, is not a mood. It is the distribution of authority.

And authority, in a classroom, is always being shaped.

Illustration in the Classroom: High School Level

Students analyze a contemporary poem layered with ambiguity and shifting perspective. The assignment is clear: develop an interpretation and defend it with textual evidence.

On paper, one student's writing is precise. Her analysis track metaphors carefully. She identifies contradictions others overlook. Her margins are filled with questions. But during discussion, her thinking recedes.

When she begins to speak, her sentences trail off midway. She restarts, then qualifies her claim before it fully forms. "I'm not sure, but maybe..." She glances sideways, scanning faces for reaction before finishing the thought.

Other students enter more decisively. Their interpretations are bold, sometimes underdeveloped, but delivered with confidence. The tempo of the conversation adjusts to their rhythm.

From the back of the room, the teacher observes the pattern.

He recognizes something uncomfortably familiar: he has often praised the students who sound certain. He has equated fluency with depth. He has allowed speed to masquerade as authority.

Later, rereading the student's written analysis, he notices that her most nuanced insights never entered the public space of the classroom. They remained on paper – thoughtful, contained.

The next week, he shifts the design. Before discussion begins, students must write a provisional claim and identify one place in the poem that unsettles it. During conversation, speakers must reference the line that complicates their interpretation, not just the one that supports it.

Certainty slows. The teacher models the process himself. "At first I thought this line suggested isolation," he says, "but now I'm reconsidering." He leaves the sentence unfinished for a moment. The pause is visible.

When the student speaks again, her tone is still measured, but she does not preemptively apologize for complexity. She traces a contradiction in the speaker's voice and allows it to stand unresolved. No one interrupts.

The shift is subtle. Some days, dominant voices still move first. Some days, she withdraws again. Belonging is not linear. But over time, she begins to inhabit the space of interpretation differently. Her posture changes less than her timing. She enters earlier. She speaks before the discussion hardens around a single reading. She stops framing her ideas as tentative by default.

Belonging here is not agreement. It is epistemic legitimacy – the understanding that one's way of reading has standing.

Artificial intelligence can generate layered literary analysis instantly. It can hold multiple interpretations in parallel without hesitation. It does not feel the risk of being wrong in front of peers. It does not negotiate the social calculus of speaking into uncertainty. Students do.

When they do not feel they belong in intellectual discourse, they do not simply speak less. They begin to believe that interpretation belongs to others – that their role is to observe, not shape. The teacher's work is not to inflate confidence. It is to redistribute authority.

Belonging, in this classroom, is not a feeling of comfort. It is the experience of being taken seriously. And that experience must be cultivated repeatedly – in who is affirmed, who is interrupted, whose uncertainty is normalized, and whose voice is allowed to carry.

The work continues. Because identity, once formed in silence, does not shift in a single conversation.

It shifts when the room changes how it listens.

Belonging is not accidental. It is constructed through relational attention, intentional norms, and the explicit affirmation of diverse voices. AI may expand access to language and information, but it cannot confer legitimacy. Teachers do that work – through noticing, inviting, and responding.

Belonging is the soil in which voice takes root. And without voice, learning remains incomplete. Belonging and voice are inseparable from meaningful learning. All levels of students engage deeply when teachers honor identity, invite perspective, and create spaces where risk-taking is safe and valued.

AI can support expression and provide scaffolds, but it cannot affirm perspective, recognize individuality, or build trust.

When teachers focus on relational scaffolding and recognition, students participate authentically, articulate reasoning, and connect learning to who they are. Belonging is not a luxury; it is the foundation for thinking, for agency, and for learning in the AI-rich age.

When teachers encourage students to articulate their thinking, even imperfectly, they communicate trust. When they honor lived experience as a source of knowledge, they expand what counts as learning. When they invite students to question dominant narratives, they foster agency.

This is especially critical in an AI age, where students may be tempted to substitute generated language for their own developing voice. The risk is not that students will sound less articulate, but that they will feel less ownership over their thinking.

Teachers counter this by valuing process over polish. They ask students to explain ideas in their own words, reflect on growth, and connect learning to lived experience. They make space for dialect, multilingualism, and alternative forms of expression. Belonging is built through these choices.

Education is about identity formation as much as knowledge acquisition. Students are learning who they are in intellectual spaces. Teachers help shape that answer. AI may assist expression, but it cannot replace the human affirmation that says, "Your thinking matters because it is yours."

When students feel they belong, they speak. When they speak, they think. When they think, they learn. Teaching voice is teaching humanity.

Invitations for Preservice Teachers

Belonging is not something students automatically feel when they enter your classroom; it is something you intentionally cultivate through tone, structure, language, and expectation.

You are invited to view every interaction as formative. The way you respond to a hesitant answer, the examples you choose, the names you learn, the stories you validate – all communicate who is legitimate in academic spaces.

Students are increasingly exposed to polished outputs that may feel distant from their lived reality. Some may internalize the idea that "correct" language is narrow or that their own voice is insufficient. Your role is to counter that narrative. When you affirm developing ideas, invite multiple perspectives, and treat identity as intellectual capital rather than a distraction, you protect students' ownership of their thinking.

You are invited to consider: Who feels most at ease speaking in your classroom – and who hesitates? Whose language patterns are affirmed, and whose are quietly corrected? When a student's voice wavers, do you move on – or make room?

Belonging also requires courage. It means interrupting dismissive comments, examining your own assumptions, and creating norms that elevate listening as much as speaking. It means recognizing that quiet students may be guarding their voice until they know it will be safe.

As you begin your career, remember this: students are not just learning your content. They are learning whether they are allowed to exist fully within it.

In a world of generated language and automated expression, your affirmation of imperfect, human thinking becomes revolutionary. Your classroom can be the place where students discover that their voice is not something to replace – but something to refine, strengthen, and trust.

Invitations for Administrators

Belonging is a foundational infrastructure for learning. It is not an initiative to layer on top of instruction; it is the condition that makes instruction possible. Policies, tracking systems, disciplinary practices, grading structures, and curriculum adoption all send signals about who belongs and whose voice counts.

In an AI-influenced landscape, the temptation may be to prioritize efficiency, innovation, and measurable outputs. Yet belonging cannot be automated or scaled through software. It is shaped through culture. Leaders influence that culture profoundly.

When administrators ensure that representation is visible in curriculum, that multilingualism is honored, that restorative practices replace exclusionary ones, and that professional development centers relational pedagogy, they reinforce legitimacy for every student.

You are invited to consider: What do your systems communicate about voice and value? Who is most affirmed by your current structures – and who must adapt to fit them? When difficult trade-offs arise, does belonging remain central or become negotiable?

Leadership also means modeling belonging for staff. Teachers who feel seen, trusted, and valued are more likely to extend that same posture to students. When administrators create space for dialogue, dissent, and reflective practice, they cultivate a professional culture grounded in voice and respect.

In an AI era, where technology may standardize expression, leaders must protect individuality. Schools that center belonging are not only more equitable – they are more intellectually vibrant. Students who feel legitimate contribute ideas. Teachers who feel trusted innovate thoughtfully. Belonging strengthens engagement, trust, and collective purpose. It is not a soft goal. It is strategic and essential.

Invitations for Homeschool Parents

You are invited to center belonging as a daily practice.

Your home learning environment can communicate, consistently and clearly, that your student's identity, questions, and perspectives matter.

As AI tools become more common, students may compare their thinking to perfectly structured outputs or standardized expectations. It is easy for a student to believe that intelligence sounds a certain way or looks a certain way. Your role is to affirm that learning is personal, developmental, and deeply human.

When you invite your student to explain ideas in their own words, when you honor cultural traditions as sources of knowledge, when you welcome curiosity – even when it disrupts a plan – you reinforce belonging. You show that education is not about conformity to an external model but contribution from within.

You are invited to consider: What signals does your home send about whose ideas matter? When your student struggles, do they feel evaluated – or supported? Are curiosity and identity treated as distractions, or as strengths?

Belonging at home also means modeling humility and growth. When you admit you are learning too, when you listen before correcting, when you celebrate effort over polish, you communicate that voice develops through relationship.

In a world saturated with information, belonging anchors identity. It tells a student, "You are not competing with machines. You are becoming yourself." That assurance nurtures confidence, resilience, and agency that extend far beyond academics.

Reflection Questions

Chapter 16: Belonging, Identity, and Voice

1. Who is most centered in your classroom culture – and who must adapt to belong?

2. Whose voices shape the intellectual direction of your classroom – and whose contributions remain peripheral?

3. When have you witnessed a student find their voice in a way that altered their sense of self?

4. How do you discern whether silence in your classroom reflects reflection, fear, or exclusion?

5. What identities are students rehearsing – or resisting – when they speak in your classroom?

Chapter 17: Creativity Is Not a Luxury

Before students can create, they must first know that their voices are legitimate.

When learners experience belonging, they are more willing to take intellectual risks, offer imperfect ideas, and experiment with possibilities. The security of being seen and valued becomes the foundation for originality. Without that foundation, creativity contracts; and with that foundation in place, imagination expands.

It is from that space of legitimacy and trust that creativity becomes possible.

In schools, creativity often appears at the margins – art on Fridays, projects after the "real work" is done, enrichment for students who finish early. When time is tight or pressure is high, creativity is the first thing to be trimmed, as though it were decorative rather than essential.

Artificial intelligence exposes the cost of that thinking.

When machines can generate text, images, music, and solutions at remarkable speed, creativity can no longer be defined as producing something novel on demand. If output alone were the measure, AI would already surpass us. What matters now is not production, but origination – how ideas are formed, combined, challenged, and reimagined.

Creativity is not about decoration. It is about problem-solving.

Teachers see this daily. The most creative students are not always the most fluent or efficient. They are the ones who make unexpected connections, question assumptions, and persist when the path forward is unclear. They tolerate ambiguity. They revise willingly. They approach constraints not as barriers, but as invitations.

Creativity shows up not only in what students produce, but in how they adapt when the predictable path fails.

AI can remix existing ideas with astonishing fluency, but it does not experience insight. It does not feel stuck. It does not take a personal risk. It does not draw from memory, culture, longing, frustration, or hope. It does not wrestle with uncertainty or sit with tension until something new emerges.

I have sat in that tension.

There are pages in this manuscript that were written and rewritten until the language felt thin. Paragraphs that sounded polished but hollow. Ideas that arrived quickly but did not hold under scrutiny. More than once, I closed the document and wondered whether fluency had replaced honesty.

It would have been easy to accept the first coherent draft. The structure was there. The sentences flowed. But something essential was missing – the friction that reveals whether an idea has been tested rather than assembled.

Creativity rarely announces itself as inspiration. More often, it feels like interruption. A line that refuses to settle. A metaphor that collapses under its own neatness. A paragraph that reads smoothly but says nothing new. Sitting with that discomfort is not efficient. It is not glamorous. It does not produce immediate output.

It produces depth.

In those moments, I learned to pause rather than polish. To ask what question I was avoiding. To follow the thread that felt unresolved rather than the one that sounded impressive. The breakthrough rarely arrived in a burst. It emerged gradually – through revision, reconsideration, and the willingness to discard what was merely competent.

Artificial intelligence can generate fluent drafts in seconds. It can suggest structures and refine phrasing with remarkable speed. But it does not feel the subtle dissatisfaction that signals an idea has not yet earned its clarity. It does not wrestle with whether a sentence is true, only whether it is coherent.

Human creativity is slower because it is personal.

It asks something of the creator.

And that asking – that quiet insistence that the work become more honest, more layered, more lived – is the place where originality begins.

That pattern is not unique to writing. It is the architecture of creativity itself.

Human creativity is born in that wrestling. It grows out of curiosity and constraint. It strengthens through experimentation and failure. Teaching creativity means protecting space for iteration and revision. It means valuing questions that do not yet have answers. It means designing tasks that require possibility-making rather than model-following.

This becomes especially urgent in an AI age, where efficiency is easily mistaken for intelligence.

Students may learn to produce quickly, but speed does not equal imagination. Creativity slows learning down in productive ways. It demands reflection, recombination, and risk. "Creativity requires time not only for invention but for incubation, reflection, and the recombination of ideas – conditions that are antithetical to speed alone" (Lang & Metcalf, 2021, p. 38). When learners are rushed from task to task or from answer to answer, they may appear fluent without ever having touched the deeper textures of innovation.

Teachers cultivate creativity when they ask students to explain how they arrived at an idea, not just present the idea itself. They invite alternatives: What else might work? What if we tried it differently? They normalize revision as a sign of depth rather than deficiency. They model curiosity and play. They demonstrate that learning is not linear, and that uncertainty is not something to escape, but something to engage.

Creativity is also deeply connected to identity.

When students are invited to bring personal experience, culture, humor, language, and perspective into their work, originality flourishes. AI cannot replicate this because it does not have a life to draw from. It cannot remember a grandmother's story, a neighborhood sound, a childhood question, or a personal failure that reshaped understanding.

Protecting creativity is not indulgence. It is preparation.

In a world where machines can generate, humans must imagine. In a landscape saturated with answers, we must cultivate possibility.

Protecting creativity is protecting humanity.

Illustration in the Classroom: Elementary Level

Students are asked to design a playground for an empty lot in their neighborhood. The parameters are simple: include at least three structures and explain why they matter.

Most students begin with familiar shapes – swings, slides, climbing walls. The drawings are colorful and symmetrical. Explanations are brief: "It would be fun." "Kids would like it." One student draws only a bench. When the teacher kneels beside her desk, she explains that her grandmother cannot climb ladders or run. "She would just watch," the student says. The teacher feels the instinct to guide her toward something more "complete." The assignment requires three structures. The bench alone does not meet the criteria. She pauses.

Later that afternoon, reviewing the plans, she notices how many designs look similar – efficient, cheerful, interchangeable. Creativity, she realizes, has drifted toward compliance. The next day, she adjusts the question. "Who might this playground leave out?" she asks. Students return to their drafts. Some add ramps. Others draw shaded areas. The student with the bench adds a wide path leading to it. She sketches flowers nearby and labels them "for sitting." The work remains uneven. Some students cling to symmetrical designs. Others erase and redraw repeatedly. One student abandons his first draft entirely and starts over. There is no moment of collective revelation. The teacher resists showcasing a single "best" design. Instead, she asks students to explain what problem their revision solved.

Over time, the plans grow less polished and more thoughtful. Structures shift in response to imagined users. Explanations lengthen. Students begin to ask one another questions about access, space, and safety. Creativity here is not decoration. It is perspective-taking under constraint.

Artificial intelligence can generate endless playground blueprints in seconds. It can optimize layouts for safety and efficiency. When creativity is reduced to novelty, it becomes surface variation. When it is cultivated as design thinking, it becomes moral imagination.

Illustration in the Classroom: Middle School Level

Students are tasked with design. They are creating a simple water filtration system using limited materials: sand, gravel, cloth, and plastic bottles. The first prototypes look nearly identical. Students follow a demonstration model closely. When tested, most filters perform similarly – adequate, but unremarkable. One group's filter fails completely. The water runs through unchanged. They look embarrassed.

The teacher senses the room tightening around success and failure. Efficiency tempts him – to show the "correct" layering, to prevent wasted time. Instead, he asks each group to document what they assumed would work. The failing group explains their reasoning. They layered sand first, believing finer material should catch debris immediately. The logic is not careless. It is incomplete. Rather than correcting them, the teacher asks: "What would happen if you reversed it?"

The redesign is slow. Some groups resist altering their working models. The failing group tests multiple versions. One improves slightly. Another collapses entirely. Frustration rises and falls. Over several class periods, variation emerges. No single design dominates. Students begin comparing not just outcomes, but reasoning. Why did that configuration work better? What trade-offs were introduced?

Creativity here is not sudden brilliance. It is iterative adjustment under constraint. Artificial intelligence can simulate thousands of filtration models instantly and identify the most efficient configuration. It does not experience uncertainty. It does not negotiate peer skepticism. It does not defend a design choice aloud. Students do.

When creativity is treated as talent, only a few risk it. When it is treated as disciplined iteration, more students remain in the process long enough to improve. The prototypes remain imperfect. But the students' willingness to revise increases.

And revision, not inspiration, is where creative expertise grows.

Illustration in the Classroom: High School Level

Students are asked to reinterpret a self-portrait using abstraction rather than realism. One student, technically skilled, produces a precise rendering on the first attempt. The proportions are accurate. The shading is controlled. It is also familiar.

When invited to distort perspective or alter color to communicate mood, she hesitates. "I don't want it to look bad," she says. The teacher recognizes the pattern – not a lack of ability, but an allegiance to correctness. He feels the temptation to model a bold version for her, to demonstrate what risk looks like. Instead, he assigns a series of constraints. Create three versions using only one color. Then three using only line. Then one that intentionally exaggerates a feature. The first attempts are cautious. The exaggerations are minimal. The distortions polite. He does not praise the safest version. He asks which felt most uncomfortable.

Over weeks, the student oscillates. Some days she returns to realism. Other days she pushes further – elongating shadows, fragmenting outlines, allowing asymmetry. There is no singular breakthrough canvas. No dramatic unveiling. But something shifts in her posture during critique. She begins to speak about intention rather than accuracy. She references choices rather than outcomes. She defends distortion as deliberate.

Creativity here is not self-expression alone. It is the capacity to remain generative when certainty is unavailable.

Artificial intelligence can produce infinite variations of self-portraits in any style. It can abstract, distort, re-render with mathematical precision. It does not fear exposure. Students do.

When creativity is framed as perfection, risk narrows. When it is framed as iterative inquiry, possibility widens. The work remains uneven. But the student returns to the canvas differently now – not asking whether it looks correct, but whether it communicates what she intended.

That question will outlast this assignment. And that is the point.

Creativity is not about producing something impressive – it is about engaging deeply with uncertainty, constraint, and possibility.

AI can generate models, optimize structures, and polish prose. It cannot cultivate the insight that emerges from struggle, nor the identity that infuses work with meaning.

Teachers remain central to this work because creativity is relational. It grows where experimentation is protected, revision is normalized, and perspective is honored. In an AI age, protecting creativity is not enrichment – it is preparation for thinking in a world where originality, not output, defines what it means to be human.

Creativity is inseparable from human guidance, reflection, and relational scaffolding.

Students engage more meaningfully when teachers protect space for experimentation, model iterative thinking, and honor personal perspective. Engagement deepens when learners perceive "opportunities for autonomy, competence, and personal relevance in their learning" (Schunk et al., 2008, p. 338). AI can generate content rapidly, but it cannot model insight, wrestle with uncertainty, or connect learning to lived experience. By prioritizing originality of thought over polished outputs, teachers nurture thinkers capable of adapting, innovating, and responding to complexity.

In an AI-rich world, protecting creativity is not a luxury; it is essential for human learning, problem-solving, and agency.

Creativity allows individuals to adapt, innovate, and respond to complexity. It prepares students for futures that cannot be predicted.

Teachers are not responsible for producing artists. They are responsible for nurturing thinkers who can imagine possibilities beyond existing systems.

AI may accelerate production, but creativity gives learning purpose.

Invitations for Preservice Teachers

Creativity is often treated as an enrichment activity. Something added after the "real work" is complete. However, creativity is not separate from rigor; it is how rigor becomes meaningful.

When replication is effortless and polished outputs can be generated in seconds, creativity becomes one of the clearest expressions of human intelligence. What distinguishes your classroom will not be how efficiently ideas are produced, but how courageously ideas are generated.

You may feel pressure to prioritize coverage, pacing, and measurable outcomes. Yet the habits that endure – adaptability, synthesis, innovation – are born in moments where students are invited to imagine alternatives, test hypotheses, and revise imperfect drafts. When you design assignments that reward experimentation over imitation, you communicate that thinking matters more than speed. When you normalize revision, you teach resilience. When you invite multiple solutions, you affirm intellectual diversity.

You are invited to consider: Where in your lessons do students have space to imagine rather than replicate? Do your assessments reward safe answers or original thought? When a student's idea diverges from the expected path, do you redirect – or explore?

Creativity also protects student agency. If AI can draft a paragraph or suggest a design, students may be tempted to substitute generation for imagination. Your role is to help them wrestle with uncertainty long enough for insight to emerge. Ask them to explain why they chose a particular path and combined ideas in unexpected ways. Celebrate risk-taking even when it leads to failure. In doing so, you help students see themselves not as consumers of solutions, but as creators of meaning.

Creativity is not a break from learning. It is the engine that makes learning transferable. And in a world where tools evolve rapidly, the ability to imagine, adapt, and innovate will outlast any specific technology.

Invitations for Administrators

You are invited to examine whether systems truly reflect the value placed on creativity. When schedules are rigid, assessments are narrow, and pacing guides leave little room for exploration, creativity can become the first casualty of efficiency. Yet in an AI era – where routine tasks are increasingly automated – human creativity is not optional. It is a strategic necessity.

Administrators shape whether creativity thrives or withers. Protecting interdisciplinary projects, allowing space for revision and iteration, and recognizing innovation in teacher evaluation all signal that imagination matters.

Professional development that focuses solely on tool mastery may create technical proficiency, but professional learning that centers problem-solving and design cultivates sustainable innovation.

You are invited to consider: Do your structures make room for experimentation, or quietly reward predictability? When teachers take intellectual risks, are they supported – or subtly redirected toward compliance? What would shift if creativity were treated not as enrichment, but as essential infrastructure?

Leadership also plays a critical role in reframing failure. If schools punish mistakes or equate deviation with weakness, creativity retreats. When leaders model curiosity, celebrate thoughtful risk-taking, and treat experimentation as learning, they create a culture where originality feels safe. In such environments, teachers feel empowered to design experiences that stretch thinking rather than simply deliver content.

Creativity prepares students not just for emerging technologies, but for complex, unpredictable futures. Administrators who protect it are not indulging a luxury – they are investing in resilience, adaptability, and the long-term vitality of their schools.

Invitations for Homeschool Parents

Home naturally offers fertile ground for creativity.

Without rigid bells or standardized pacing, you can follow curiosity where it leads. Creativity flourishes when students are given time to play with ideas, experiment with materials, and revise without fear of immediate evaluation.

As AI tools become more accessible, it can be tempting to prioritize efficiency or polished outcomes. A beautifully formatted essay or instantly generated illustration may feel satisfying. But creativity grows in the messy middle – the drafting, questioning, revising, and reimagining. When you encourage your student to try again, explore alternatives, or combine ideas from different experiences, you nurture confidence and ownership.

You are invited to consider: Where in your home does creativity have room to unfold slowly? When a project becomes imperfect or frustrating, do you step in to fix it – or create space for persistence? Are outcomes celebrated more than process, or is experimentation honored in its own right?

Creativity is not about innate talent. It is about persistence, courage, and curiosity. When students learn that they can generate ideas, reshape problems, and express their unique perspectives, they develop a sense of agency that no algorithm can provide.

In honoring the creative process, you teach your student something enduring: that innovation comes from human imagination, and that their voice and ideas matter.

In a rapidly changing world, that lesson may be one of the most powerful gifts you can offer.

Reflection Questions

Chapter 17: Creativity Is Not a Luxury

1. If your classroom eliminated all creative thinking, what would students lose beyond engagement?

2. Where in your classroom are students required to generate something that does not already exist?

3. When have you felt the thrill of creating something new in your own thinking – and where do students experience that same possibility?

4. What consequences – subtle or explicit – discourage students from taking creative risks in your classroom?

5. If your students encounter a problem no one has solved yet, how prepared are they to respond imaginatively rather than procedurally?

PART VI — PRACTICAL WISDOM FOR REAL CLASSROOMS

Chapter 18: What to Do Monday Morning

Creativity must be protected. Belonging must be constructed. Attention must be cultivated. Ethics must be modeled. Influence must be earned. None of these are add-ons; they are the architecture of meaningful learning.

Which leads to the natural questions: How does this translate into my classroom, my syllabus, my students, tomorrow morning? The shift required in an AI age is not about reinventing teaching. It is about realigning it. It is about stripping away what was never essential and clarifying what always was. The goal is not to add more tools, more surveillance, more rules, or more professional obligations. The goal is to return to purpose and design learning intentionally around it. This work is not theoretical. It is practical. And it can begin immediately.

For me, one Monday morning, the rules arrived before the students did.

In a residential classroom, policy is not abstract. It is posted. Practiced. Enforced. There are protocols for movement, materials, proximity, and tone. There are adults in the room who do not belong to you – staff whose job is safety, compliance, documentation. Even when they are not evaluating you, they feel like evaluators. Their presence changes your breathing. It changes your risk tolerance. It changes what you dare to try.

A student entered late, jaw set, eyes scanning for threat or exit. The script in my head was clear: redirect, document, maintain order. The other script – the one that mattered – whispered something else: Ask what happened. Make room. Don't reduce him to a behavior to be managed. I felt torn in the space between those scripts.

If I held firm to the rules, I could keep the room predictable and keep my job. If I leaned toward connection, I could honor the student's humanity – but I risked appearing "soft" or careless, or even out of compliance. Safety was real. So was dignity. And in that room, both had consequences. So, I chose a third way: I kept the boundary and widened the entry.

I quietly said "You're here. Take your seat. We'll start." Then I added, "When you're ready, I want to hear what today has been like." The room stayed safe. The student stayed seen. And I learned that purpose can lead even inside policy – that you can hold the line without hardening your heart.

Begin with Purpose, not Policy

When confronted with new technology, the instinct is often to draft rules. Schools publish acceptable-use statements. Teachers outline restrictions. Syllabi expand with disclaimers and consequences. But policy without purpose breeds compliance at best – and resistance at worst.

Instead of opening the year with a list of prohibitions, begin with a conversation. Ask students what learning is for. Ask them what helps them understand something deeply and what simply helps them finish faster. Invite them to reflect on moments when they truly learned – and what conditions made that possible.

Then introduce AI within that frame. Use that frame to guide recommended practices and share those with people making policies. Policy makers in schools are often removed from the classroom and your advocacy here, focused on what honors students and integrity, can hold weight.

Position AI not as a threat to integrity, but as a tool that requires judgment. Discuss when tools support learning and when they replace it. Ask students: When does AI help you think? When does it do the thinking for you? How do you know the difference?

Students are far more likely to act ethically when they understand the *why* behind expectations. When the purpose of learning is clear – growth, reasoning, reflection – boundaries make sense. They are no longer arbitrary restrictions; they are aligned with shared goals.

Purpose builds trust. Policy alone enforces control.

In an AI age, clarity of purpose becomes the anchor. If students understand that the aim of school is not performance but development, not product but process, they are more likely to approach tools thoughtfully.

Start there. Everything else becomes easier to design.

Illustration in the Classroom: Elementary Level

On the first week of school, a teacher gathers her students in a circle. No devices are present yet. No rules have been reviewed. She asks a simple question. "What makes learning exciting?" Hands rise quickly. "When we get to choose." "When we try again." "When we figure something out ourselves."

She writes their words on chart paper. She circles phrases that repeat "try again", "figure out", "choose". Then she asks, "If we had a tool that could give us answers really fast, how should we use it so we still get to do those things?"

The room grows thoughtful. "If it gives us ideas, but we still pick." "If it helps with hard words." "If we don't let it do everything."

Weeks later, when students begin a science project on insects, she allows them to use AI to generate possible species to research. The list appears quickly. But she does not praise speed. Instead, she asks, "Which one makes you curious?"

Students circle their selections and explain why to a partner. Some change their minds after talking. The decision becomes the work.

Artificial intelligence can produce options instantly. It cannot decide which question feels alive to a child.

Because purpose was named first, the tool does not become the center of the project. Curiosity does.

The lesson did not begin with restriction. It began with intention.

Illustration in the Classroom: Middle School Level

At the start of a unit on ecosystems, a teacher writes a question on the board. "Why do scientists test things themselves if the answers are already published?" Students hesitate. "To make sure it's right." "To understand it better." "To see it happen." He nods. "Science is not about having information. It's about making sense of it."

Only after that conversation does he introduce AI as a possible assistant – useful for organizing measurements or visualizing data. But he frames its role clearly. "It can sort patterns," he says. "It cannot notice what surprises you."

During a field study of local water quality, students input their data into an AI tool that produces clean graphs and summaries within seconds.

The visuals are polished. The teacher waits. "What did you notice before the graph?" he asks. Students return to their field notes – the uneven temperature reading, the smell near the shoreline, the algae clumped unexpectedly in one corner. The summary did not include those details.

Artificial intelligence can condense information efficiently. It does not experience observation. It does not wrestle with anomaly.

Because the purpose of science was named before the tool appeared, the class understands the distinction. The graph is helpful. It is not authoritative. Understanding remains theirs to construct.

Illustration in the Classroom: High School Level

A teacher writes: "What does it mean to think like a historian?" Students respond: analyzing sources, identifying bias, understanding context. She listens and then asks: "If a tool summarizes a speech perfectly, have you done the work of a historian?" Silence.

Rather than issuing warnings about misuse, she proposes a comparison. Students input a primary source into an AI tool and examine the summary it produces.

It is accurate. Efficient. Organized. Then she asks, "What does it not wrestle with?" Students notice what the summary flattens – the urgency in the language, the tension of the moment, the moral stakes.

The teacher does not condemn the tool. She clarifies the role. "AI can organize information," she says. "It cannot decide what matters historically."

Together, the class drafts principles for tool use. AI may clarify language. It may surface patterns. It may assist with organization. It may not substitute for interpretation.

Because the purpose of historical thinking was named first – discernment, context, moral reasoning – the boundaries feel coherent rather than imposed.

The policy follows the purpose. And when purpose is clear, tools remain in proportion.

Make Thinking Visible

Design tasks that require explanation, reflection, and revision. Ask students to show how they arrived at an answer, not just what the answer is. Use prompts like "What changed in your thinking?" or "Why did you choose this approach?"

When thinking is visible, learning becomes harder to outsource. In practical terms, this means shifting assignments from product-only submissions to process-rich experiences.

Instead of collecting only a polished final essay, ask students to submit annotated drafts that reveal their thinking in motion – highlighting sentences they revised, noting where an argument felt unclear, and explaining why they reorganized a paragraph or replaced a claim. Invite them to attach a brief reflection describing what feedback influenced their revisions and where they are still uncertain.

In mathematics, move beyond grading the final answer by requiring students to show multiple solution paths, record a short audio explanation of their reasoning, or identify where they initially misapplied a formula and how they corrected it. Ask them to label productive mistakes and explain what those detours clarified.

In science, have students write predictions before conducting an experiment, then compare those predictions to actual results and analyze why discrepancies occurred – was the variable misunderstood, the measurement flawed, the assumption incomplete?

In social studies, provide competing primary sources and require students to justify how they weighed credibility, context, and bias, citing specific language that shaped their interpretation.

In literature, ask students to defend an interpretation in writing and then revise it after peer dialogue, noting how their thinking evolved. These structures reposition thinking from a hidden internal process to observable evidence. Reasoning becomes the assignment, not an afterthought.

By designing tasks that require explanation, revision, and comparison, teachers ensure that understanding is demonstrated through process, not merely product. Reflection questions should not feel like add-ons; they should be embedded into the design.

A history paper might conclude with, "Which source most influenced your interpretation, and why?" – requiring students to identify not only the evidence they used, but how it shaped their reasoning. They might quote the passage that shifted their perspective and explain whether that influence strengthened, complicated, or unsettled their original claim. Rather than listing sources as support, they are asked to trace the evolution of their thinking and make their interpretive moves visible.

A lab report might include, "If you repeated this experiment, what would you adjust and why?" – prompting students to reflect on procedural choices, measurement error, or assumptions that influenced outcomes. Instead of treating results as final, they analyze limitations, reconsider variables, and propose informed revisions. The focus shifts from reporting findings to examining how those findings were constructed.

A creative project might ask, "What risks did you take, and what did you learn from them?" – encouraging students to name moments of uncertainty, experimentation, or failure that contributed to growth. They might describe a design choice that did not work as expected and explain how revising it clarified their intention or strengthened their vision. Creativity becomes less about polish and more about process, iteration, and reflection.

A problem set could end with, "Where did your strategy shift, and what prompted that change?" – inviting students to narrate the evolution of their reasoning. By identifying where an assumption failed, a diagram clarified confusion, or a new approach emerged, students transform mathematics from silent procedure into visible thought. The emphasis shifts from speed and accuracy to discernment, flexibility, and the willingness to revise under new insight.

A discussion response might ask, "How did your interpretation evolve after hearing your peers?" – requiring students to trace intellectual movement rather than defend a static position. They might identify a comment that complicated their claim or a line of text they initially overlooked. Literature becomes less about arriving at the 'right' reading and more about cultivating responsiveness, depth, and revision in community.

A document-based response might ask, "How did your interpretation change as you evaluated additional sources, and what criteria guided that shift?" – requiring students to make visible the reasoning behind their conclusions rather than presenting a single polished claim. They might identify language that revealed bias, contextual factors that reshaped credibility, or tensions between competing accounts that complicated certainty. Instead of treating evidence as interchangeable support, students show how they weighed reliability, motive, and limitation.

These prompts move reflection from optional to essential. They communicate that learning is not measured solely by completion or correctness, but by revision, discernment, and intellectual risk-taking.

By embedding these questions into assignments, teachers make visible the ongoing process of thinking – signaling that learning is iterative and thoughtful, not merely efficient.

Revision, too, must be normalized. When students understand that drafts are expected to change, they begin to see growth as evidence of learning rather than proof of inadequacy. Structured revision cycles – peer feedback, conferences, reflection journals – communicate that ideas are meant to evolve. AI can generate a draft, but it cannot engage in authentic self-evaluation unless prompted by a human who asks, "Does this reflect what you actually think?"

Making thinking visible also reshapes classroom culture. Students begin to listen differently when peers explain reasoning rather than announce answers. Dialogue becomes richer when disagreement is explored through questions rather than corrected with authority or assigned right or wrong. The emphasis moves from correctness to clarity, from speed to depth.

In an AI age, where polished outputs are easy to produce, the real work of education lies in cultivating metacognition – the ability to notice one's own thinking. When teachers consistently ask students to explain, reflect, and revise, they are not adding extra steps; they are strengthening the foundation of learning itself.

Visible thinking fosters ownership. Ownership fosters integrity. And integrity makes learning both durable and deeply human.

Illustration in the Classroom: Elementary level

A student solves a word problem and writes the correct number at the bottom of the page. The answer is accurate. The reasoning is invisible.

The teacher adjusts the expectation. "Show me how you decided," she says.

Students draw diagrams. They write number sentences. They explain to partners why they grouped objects the way they did.

The teacher models uncertainty. She sketches one approach, pauses, crosses it out. "I'm not convinced yet," she says. "Let me try another way."

Confusion becomes visible – and permitted.

Over time, the student begins explaining not only what she did, but why she did it. She notices that different strategies can reach the same solution. She compares methods instead of defending one reflexively.

Correctness still matters. But clarity of reasoning becomes central.

The number is no longer the assignment. The thinking is.

Illustration in the Classroom: Middle School Level

Students calculate density accurately. The table is neat.
The formula applied correctly. Previously, that would have
been enough. This time, the teacher asks students to annotate their reports. Where did you hesitate? What assumption did you make? How might that choice affect your conclusion?

The student notices a slight inconsistency in one measurement. He recalls rushing the reading. He recalculates and explains how the shift alters interpretation.

The focus moves from correct computation to methodological awareness.

Artificial intelligence could perform the calculations instantly. It could even suggest likely errors. It cannot surface the student's reasoning unless he articulates it.

The lab becomes less about arriving at the expected value and more about understanding how conclusions are constructed.

Scientific thinking becomes visible.

Illustration in the Classroom: High School Level

A student offers an interpretation of a poem's central tension. Her explanation is thoughtful but concise. In another setting, it might be affirmed quickly – or replaced with a more polished reading. Instead, the teacher pauses. "Map your reasoning," she says.

The student moves to the board. She writes her claim at the top and begins drawing lines to specific phrases in the poem. She circles imagery that shaped her interpretation. She annotates a word whose connotation altered her initial assumption. The room grows quieter – not because the answer is correct, but because the thinking is unfolding. A peer raises a question. "What if that line suggests regret instead of defiance?"

The student returns to the text. She rereads the stanza aloud. Her marker hovers for a moment before she adds another arrow, this time pointing to a different phrase. She revises her explanation slightly – not abandoning her claim, but refining it.

The teacher does not intervene to resolve the tension. She lets the class sit with the ambiguity. Another student points out a contradiction between two lines. The original speaker responds by acknowledging that she had overlooked it. She adjusts her map again, layering complexity rather than retreating from it.

Nothing dramatic occurs. No breakthrough declaration. No applause. But the class witnesses something rare: interpretation as construction rather than performance. The teacher steps back. "What shifted?" she asks. The student answers carefully. "I thought it was about anger. Now I think it's more complicated." The poem has not changed. Her reasoning has.

Visible thinking requires vulnerability. It requires the willingness to let others see how conclusions are formed. When students are asked to make their reasoning observable – to trace claims, acknowledge assumptions, revise under scrutiny – analysis becomes more than delivery. It becomes ownership. And ownership is harder to outsource than answers.

Making thinking visible shifts learning from answer production to meaning construction. When students are asked to explain, reflect, and revise publicly and thoughtfully, they develop habits of mind that extend beyond any single assignment – or any single tool.

Normalize Revision

Build in intentional opportunities for students to revise their work based on feedback. Make revision visible, structured, and expected – not an optional add-on for those who didn't get it right the first time. When drafts are treated as a natural stage of development rather than a sign of deficiency, students begin to understand that strong thinking rarely appears fully formed. Normalize language such as, "This is your first attempt," or "Let's see how your thinking evolves."

Provide time for students to revisit ideas, strengthen reasoning, and clarify misunderstandings. When revision is embedded into the design of learning, it communicates something powerful: that growth matters more than speed, and that understanding deepens through reflection. Revision is where ownership develops. It is where students move from repeating ideas to reshaping them.

Teachers can support this by asking metacognitive prompts: "What changed in your thinking after receiving feedback?" "What part of your argument feels stronger now – and why?" "Where did you struggle, and what did you try next?" These questions reinforce that learning is not a straight line. It bends, pauses, recalibrates, and strengthens over time.

When revision is normalized, classrooms shift from performance culture to growth culture. Mistakes become information. Feedback becomes fuel. Students begin to see drafts not as evidence of inadequacy, but as evidence of effort and engagement. They learn that insight often emerges in the second or third attempt –rather than the first. Revision communicates a deeper truth about learning: it is interactive. It requires humility, persistence, and the willingness to rethink. In modeling and protecting space for revision, teachers teach far more than content – they teach resilience, reflection, and intellectual courage.

Illustration in the Classroom: Elementary Level

A student finishes his first draft of a short story. He rereads the final paragraph, hesitates, and crumples the page. "It's wrong," he says quietly. The teacher has seen this before – the belief that first attempts should arrive complete. She recognizes the impulse in herself as well: the desire for work to look finished quickly, to signal progress. Instead of asking him to "try again," she shifts the frame. "Let me show you something," she says.

She brings out her own draft – marked with arrows, crossed-out phrases, added sentences written into the margins. She does not present it as a model. She presents it as a process. "This was my first version," she explains. "I thought I was done." She reads one sentence aloud and pauses. "Now it sounds flat to me. I'm going to move it." She revises in front of him. Not dramatically. Not performatively. Simply. The page becomes a record of thinking rather than a verdict on quality. She does not ask him to rewrite the whole story. She narrows the task. "Choose one paragraph," she says. "Make it clearer." The scope shrinks. The overwhelm loosens.

He rereads and adds a line of dialogue. Then he inserts a sensory detail he had imagined but not written. The paragraph strengthens incrementally. Nothing transforms instantly. The story is still uneven. Some sentences remain awkward. But the paper is no longer in the trash.

Over the following weeks, revision becomes routine rather than rescue. Students use colored pencils to mark changes between drafts. They compare versions and describe what shifted. They begin to expect that ideas will evolve. The teacher notices a subtle change. Students no longer ask, "Is this right?" as often. They ask, "Is this clearer?" Revision is not about correcting error. It is about strengthening meaning.

By the end of the unit, the student places his first and later drafts side by side. The improvement is visible, but more importantly, the process is understood. The first attempt was not wrong. It was incomplete. And incompleteness, in this classroom, is expected.

Illustration in the Classroom: Middle School Level

A student submits a draft essay on the Industrial Revolution. The structure is clear. The events are described accurately. The timeline is orderly. On the surface, it appears complete. But most of the paragraphs summarize rather than interpret. The causes are listed. The consequences are mentioned. The reasoning remains implicit.

In the past, the teacher might have assigned a grade and moved forward. Instead, he responds with questions in the margins. Why did factory conditions lead to reform? Whose perspective is missing here? Where does your interpretation become visible? He circles one paragraph and writes, "This is where your thinking could go deeper." When the draft is returned, the student feels momentarily unsettled. She believed she was finished. The structure felt solid.

During a revision workshop, the teacher models the shift from summary to analysis using a sample paragraph. He adds a causal phrase. He introduces a counterpoint. He demonstrates how a sentence can move from description to interpretation with a few precise words. He does not dramatize the change. He shows it. The student rereads her draft. She reorganizes one section to contrast economic growth with labor conditions. She inserts a paragraph considering how a factory owner might have justified practices that workers resisted.

In the margin of her revision, she responds to a prompt: What changed in your thinking? "I used to think industrialization was mostly positive," she writes. "Now I see it depended on who you were." The revision is not cosmetic. It is conceptual.

Through revision, the student begins to understand that a draft is not a performance of knowledge. It is a position to be reconsidered. The final paper is stronger. But more importantly, her reasoning is more layered. Revision shifts from correction to recalibration.

And recalibration becomes part of how she approaches new questions.

Illustration in the Classroom: High School Level

A student drafts a lab report analyzing enzyme activity under varying pH levels. The calculations are accurate. The graph is labeled correctly. The formatting is clean. On the surface, the report appears strong. But the explanations are compressed. The optimal pH is identified without examining the assumptions beneath the claim. One trial deviates from the expected pattern, yet the discrepancy is mentioned only briefly. The numbers are correct. The reasoning is abbreviated.

Instead of assigning a final grade, the teacher writes in the margin: What assumptions are built into this conclusion? What variables might have influenced this outcome? Where does your interpretation become visible? She does not rewrite the section for him. She asks him to walk through his reasoning during a brief conference.

As he explains the procedure aloud, he begins to notice what he had taken for granted – that temperature remained constant, that concentrations were precise, that measurement timing was exact. He recognizes that these conditions were approximated, not controlled perfectly. The data did not change. His understanding of the data does.

He revises the report. He adds a paragraph analyzing potential experimental limitations. He recalculates one data set after reconsidering measurement precision. He clarifies how small procedural differences might account for variation in enzyme activity. In his reflection, he writes: "My first draft showed the result. This one explains how I got there." Revision, here, is not correction. It is recalibration.

Over time, the student begins anticipating revision. He drafts with the expectation that interpretation will evolve. He asks sharper questions earlier in the process. The report improves. But more importantly, his scientific thinking becomes iterative rather than declarative.

And iteration, not immediacy, is what sustains expertise.

Use AI Transparently

Model how AI can support learning without replacing it. Do not treat it as a hidden shortcut or a forbidden tool. Bring it into the open. Project it on the screen. Think aloud as you use it. Show students not only what it produces, but how you evaluate what it produces.

Transparency shifts AI from temptation to tool. When teachers hide AI, students experiment with it in isolation. When teachers demonize AI, students use it defensively. But when teachers model thoughtful use, they transform it into an object of inquiry. With inquiry, students begin to see AI not as an authority to trust blindly, nor as a loophole to exploit, but as a system to question. For example, generate a response to a prompt in front of students and ask: "What does this answer do well?" "Where is it vague?" "What assumptions is it making?" "What might be missing?"

Let students critique tone, accuracy, bias, and depth. Invite them to compare the AI response to a primary source, their own draft, or a peer's reasoning. Ask them to revise the AI's output rather than submit it unchanged. This reframes AI from final product to first draft.

Transparency also means naming limits. Explain that AI can summarize but not experience. It can remix language but not wrestle with uncertainty. It can simulate but cannot defend its reasoning in dialogue. When students see these boundaries clearly, they are less likely to overestimate its authority.

Model ethical attribution. If AI assists with brainstorming or organization, say so openly. Demonstrate how to cite assistance when appropriate. Normalize the idea that responsible use includes acknowledgment.

Transparency builds discernment because it makes thinking visible. Students learn that the important work happens after the response is generated – in the questioning, revising, connecting, and refining. They learn that intellectual ownership comes from interpretation, not from production alone.

Ultimately, transparency communicates confidence. It says to students: "We are not afraid of this tool. We are capable of guiding it". And that message cultivates the very discernment students will need long after the classroom has changed again.

Illustration in the Classroom: Elementary Level

After finishing a short story together, a student uses an AI reading assistant to generate comprehension questions. The list appears quickly – grammatically correct, neatly phrased. On the surface, it seems usable. The teacher does not collect it quietly. She projects the questions on the board. "Let's look at these together," she says. The room reads. "Is this question asking us to think," she asks, "or just to find a sentence?"

Students hesitate. One reads the question again aloud. It asks who found the lost dog. The answer is stated plainly in the second paragraph. "That's easy," someone says. "Easy can be fine," the teacher replies. "But is it interesting?" She circles another question – factually accurate but vague. "What makes this story important?" it asks. "What would make that harder to answer?" she continues.

Hands rise. One student suggests adding, "for the main character." Another proposes asking why the ending felt surprising. The teacher does not rewrite the question herself. She invites the original student to compare the AI-generated version to a line from the text. They notice that the tool merged two characters' actions into one. The mistake is small. It is visible. The class revises one question together, tightening it so that it requires inference rather than recall. Nothing dramatic occurs. The AI output remains on the screen. So does the markup layered over it.

Over time, the student begins drafting two of his own questions before consulting the tool. He compares tone. He asks whether the AI's prompts reach the heart of the story or skim its surface. The tool remains present. It is not hidden. It is not feared. It is examined.

Transparency does not eliminate the tool's use. It clarifies its place.

In this classroom, the strongest questions are not the fastest ones. They are the ones revised through attention.

Illustration in the Classroom: Middle School Level

A student researching climate change uses AI to summarize a scientific article. The summary appears quickly – concise, organized, technically accurate. It feels efficient. The teacher does not praise the speed. He projects both the original article and the AI-generated summary side by side. "Let's compare," he says. Students begin reading quietly. Highlighters move between the two texts. "What disappeared?" he asks. At first, the room is unsure. The summary seems complete.

Then someone notices a missing paragraph about regional variability. Another points out that the original author's tone conveyed urgency, while the summary sounds neutral. A third observes that a specific data limitation mentioned in the article does not appear in the summary. The teacher underlines one omission but does not elaborate. "Why might that matter?" he asks.

Students begin tracing the implications. They mark where nuance has been compressed and where emphasis has shifted. Together, they revise one section of the summary, restoring a detail the tool had flattened.

The AI output remains visible on the screen – but so do the annotations layered over it. Artificial intelligence can condense information efficiently. It does not decide which details deserve weight. It does not feel the difference between neutrality and urgency.

Over time, the student's habits shift. She consults the original source first. She uses AI to organize notes, not replace them. She checks tone. She looks for what might be missing.

The tool remains part of the process. It is no longer mistaken for the final word. Transparency does not eliminate efficiency. It restores proportion. And in that proportion, discernment begins to take shape.

Illustration in the Classroom: High School Level

A student preparing for a policy debate. She uses AI to generate a structured outline of arguments and counterarguments. The bullet points are organized. The phrasing is persuasive. The structure appears ready for delivery. It looks complete. Instead of discouraging the tool's use, the teacher asks her to project the draft. The outline appears on the screen. "Where do these claims originate?" the teacher asks. Silence settles briefly.

Students begin reading more closely. One circles a statement that sounds definitive but lacks citation. Another notes that a complex funding issue has been reduced to a single cause. A third asks whose perspective is centered in the framing. The teacher does not supply answers. "What assumptions are embedded here?" she asks.

The student cross-checks the AI-generated points against her research articles. She discovers that one claim compresses multiple stakeholder positions into one simplified argument. She revises it, adding specificity and context. Another point is removed entirely after she cannot trace its source.

The AI draft remains visible – now layered with annotations, questions, and revisions. Artificial intelligence can assemble persuasive language efficiently. It does not evaluate local nuance. It does not assume responsibility for framing. It does not reconsider a claim unless prompted.

The revision process becomes the work.

By the time the debate begins, the student's outline looks less polished but more grounded. Citations are added. Claims are narrower. Counterarguments are more precise. The tool has not disappeared. It has been repositioned.

Transparency does not eliminate AI from the room. It clarifies that authority belongs to the thinker who questions it. And in that questioning, intellectual ownership becomes visible.

Shift Assessment Toward Reflection

If assessment is to privilege thinking over polish, it must intentionally shift toward reflection.

When we evaluate only final products, we reinforce performance. When we evaluate reasoning, revision, and critique, we cultivate judgment. In an AI-rich classroom, this shift becomes even more essential. Rather than treating artificial intelligence as a hidden shortcut or an unspoken threat, bring it into the open as an object of inquiry.

Model how it can support brainstorming, generate structure, or surface initial ideas – while also making visible its limitations. Show students what it does well – organizing information quickly, producing fluent summaries – and where it falls short – flattening nuance, overlooking context, or presenting confident language without depth. Transparency removes mystery, and when mystery disappears, so does much of the temptation to misuse.

Project an AI-generated response in front of the class and examine it together. Ask: What is accurate here? What is vague? What feels generic? What might be missing? Invite students to annotate the output, circling strong reasoning and underlining unsupported claims. Encourage them to identify where evidence is implied but not cited, where claims lack specificity, or where the argument feels formulaic. Then challenge them to improve it. Add nuance. Insert textual evidence. Clarify assumptions. Introduce counterarguments. Revise tone.

When students learn to critique, refine, and extend AI-generated material rather than copy it, they practice discernment. Assessment, then, becomes less about detecting misuse and more about evaluating the quality of thinking students demonstrate in response. Reflective analysis – "What did the AI overlook?" "How did your revision strengthen the argument?" "What decisions did you make and why?" – makes reasoning visible and positions judgment, not compliance, at the center of learning.

Model your own thinking aloud. "The AI summary is efficient, but it doesn't address the counterargument we discussed yesterday." Or, "This explanation is clear, but it assumes all readers share the same background knowledge." By narrating your evaluation process, you make judgment visible. Students begin to see that expertise lies not in generating text quickly, but in evaluating it thoughtfully.

Transparency also means naming boundaries. Be explicit about when AI is appropriate – for brainstorming, checking grammar, generating practice problems – and when it is not – for demonstrating personal reflection, practicing foundational skills, or submitting final assessments without attribution. Students are far more likely to act ethically when expectations are clear and grounded in purpose rather than fear.

Invite students into the ethical conversation rather than assuming their intentions. Ask them directly: "When does AI genuinely help you think more clearly?" "When does it begin to do the thinking for you?" Encourage them to provide specific examples – "Was it helpful in generating ideas?" "Did it organize your outline?" "Did it obscure confusion you needed to confront yourself?" Their responses often reveal far more insight than blanket suspicions about misuse.

These discussions cultivate metacognition – an awareness not only of what students think, but how tools shape their thinking. To reinforce this awareness, assess through reflection. Ask students to submit brief learning statements alongside assignments: "How did you use AI in this task?" "What decisions remained yours?" "Where did you revise or override the tool?" When students articulate their choices, reflection becomes evidence of learning, and ethical use shifts from compliance to conscious practice.

Transparency builds trust. It signals that you are not competing with technology, nor pretending it does not exist. Instead, you are guiding students in how to engage it responsibly. When AI is modeled openly – critiqued, refined, and contextualized – it becomes what it should be: a support for learning, not a substitute for it.

Illustration in the Classroom: Elementary Level

Students complete a short informational paragraph about frogs. They are invited to use an AI tool. It produces – organized, fluent, neatly structured summaries. Instead of collecting the paragraphs for grading, the teacher projects one AI-generated summary beside a student's draft. "Let's look at both," she says. The class reads quietly. "What does the AI include?" she asks. Students identify familiar facts: habitat, diet, life cycle. "What might be missing?"

A student notices that the AI summary lists stages of development but does not mention the pond they observed during last week's field walk. Another points out that the tone sounds formal and distant compared to the student's description of tadpoles "wiggling like commas." The teacher does not declare one version better. Instead, she reframes the assessment. "Your grade will include your reflection," she explains. "Tell me what you kept, what you changed, and why."

Students annotate their drafts. One writes, "I kept the life cycle order, but I added what we saw." Another notes, "The AI didn't say why frogs are important here." The focus shifts from whose paragraph is smoother to whose thinking is clearer. The AI output remains visible – not as a model, but as material to examine. In their learning journals, students respond to three prompts: How did the AI help you? What did you revise? What decisions were yours? At first, responses are brief.

Over time, they grow more specific. Students begin identifying when the tool organizes ideas efficiently and when it flattens detail. They notice that adding personal observation changes meaning. Assessment no longer measures only whether facts are correct. It evaluates discernment. The strongest submissions are not the most polished. They are the ones that show thoughtful revision and intentional choice.

Artificial intelligence may generate language quickly. Reflection reveals who is thinking. And when reflection becomes part of assessment, ownership becomes visible.

Illustration in the Classroom: Middle School Level

Students research the causes of the Great Migration.
They use AI to help organize notes or generate preliminary summaries. Several students submit well-structured essays. The writing is clear. The paragraphs transition smoothly. The structure feels polished.

Instead of grading solely on content accuracy and organization, the teacher includes a required reflection section: Where did AI assist you in this assignment? What did you revise or reject? How did your understanding shift during the process? One student writes that she used AI to outline possible causes of migration. The list was efficient – economic opportunity, labor demand, racial violence. "But it didn't explain how those things felt," she notes. In her reflection, she describes reading firsthand letters from migrants and realizing that safety and dignity shaped decisions as much as wages did. She rewrote two paragraphs after recognizing that her draft sounded detached.

The teacher responds not only to the essay, but to the reflection itself. "What prompted you to revise?" he asks in the margin. During a brief conference, she explains that the AI-generated outline helped her see structure, but the primary sources shifted her perspective. She identifies one paragraph she kept mostly intact and another she substantially reworked after rereading her sources. The assessment includes both the essay and her analysis of how it evolved.

In class, the teacher projects an anonymized reflection and asks, "What does this show about the writer's thinking?" Students begin to recognize that strong work is not just fluent – it is examined. Reflection becomes the evidence of ownership.

Over time, students begin asking themselves different questions before submitting work: What did I accept without checking? What did I revise – and why? Where did my understanding change? The shift is subtle. The essays remain structured. But assessment now values discernment as much as delivery. And discernment cannot be outsourced.

Illustration in the Classroom: High School Level

Students write position papers on a current policy issue.
They use AI for brainstorming or organizing arguments, provided they disclose how it was used. One student submits a well-structured essay. The claims are clear. Counterarguments are acknowledged. Citations are present. Alongside the paper is a required reflection: How did AI assist you in this task? Where did you revise or override its suggestions? What decisions remained fully yours? In her reflection, she explains that she used AI to generate a list of potential arguments. "It helped me see the structure," she writes. "But some claims were too broad." She describes discarding one suggested argument after discovering it oversimplified the policy's fiscal impact. She narrowed another claim after reviewing primary data. She notes that the AI-generated counterargument felt persuasive but lacked a stakeholder perspective she found important.

The teacher evaluates both the essay and the reflection. In the margin, he responds not only to her thesis, but to her analysis of the tool's influence: "What made you question that claim?" "How did reviewing the original data change your stance?"

During class, anonymized excerpts from reflections are projected. Students discuss what discernment looks like – where a writer accepted structure but rejected framing, where revision deepened nuance, where citation replaced generalization. The AI's role is neither hidden nor celebrated. It is examined. Assessment now includes the quality of the student's reasoning about the tool – not merely the fluency of the final paper.

Over time, students begin drafting with greater awareness. They ask themselves where structure ends and judgment begins. They anticipate the reflection before it is assigned. The essays remain articulate. But what is evaluated most closely is the thinking behind them – how claims were tested, where assumptions were challenged, and why revisions occurred.

When assessment privileges reflection, ownership becomes visible.

And ownership is the measure that endures.

Protect Attention

Design moments of sustained focus intentionally rather than accidentally. In a world increasingly engineered for interruption, depth must be constructed. Slow the pace of instruction so students have time to wrestle with ideas instead of rushing toward completion. Replace constant transitions with longer blocks of concentrated inquiry. Protect stretches of uninterrupted reading, writing, problem-solving, and discussion where students remain with a task long enough for complexity to surface.

Build in silence – not as punishment, but as space. Silence allows thinking to gather itself. It creates room for reflection before response, for noticing before judging, for questions to form rather than evaporate. Pair this with structured discussion that requires listening as much as speaking. When students are expected to summarize a peer's idea before offering their own, attention becomes relational rather than performative.

Create "deep work" routines that signal to students: this time is different. Devices are closed. Notifications are silenced. The expectation is not speed but presence. Over time, students begin to feel the difference between shallow engagement and sustained focus. They experience the cognitive reward of clarity that emerges only after effort.

Explain explicitly why attention matters. Students live in systems that monetize distraction; they deserve to understand the cost. Teach them that attention strengthens memory, sharpens reasoning, and builds self-regulation. Show them how fragmented focus leads to surface-level understanding, while sustained engagement leads to insight. When they recognize that attention is not control imposed from outside but a skill that expands their agency, they can begin to value it.

Students will not protect what they have never been taught to notice. When teachers model attentiveness – pausing before answering, listening fully, returning to unfinished ideas – they demonstrate that focus is not rigidity but care.

Protecting attention is not nostalgic resistance to technology; it is preparation for thoughtful living. In cultivating attention, teachers are not merely managing classrooms – they are strengthening minds.

Illustration in the Classroom: Elementary Level

Students are asked to examine a leaf with magnifying lenses and sketch what they observe. In previous years, the task moved quickly. Students labeled major parts and turned in their diagrams within minutes. This time, the teacher adjusts the structure. "You have ten quiet minutes," she says. "Your job is to notice something you didn't see at first." The room shifts. At first, pencils hover. A few students glance toward classmates. One whispers, "I'm done." The teacher does not fill the silence.

Instead, she kneels beside a desk and studies a leaf herself. She traces a vein with her finger. She redraws a section of her own sketch. "I missed this part before," she says softly. The pace slows. Students begin leaning closer to their leaves. One notices tiny irregular lines branching off the main vein. Another observes that two leaves of the same plant are not identical.

Midway through, she pauses the class. "What changed in your drawing?" she asks. Students add details. Erasers move. Lines are thickened. Labels are adjusted. Nothing dramatic occurs. No applause-worthy discovery. But the drawings grow more precise.

Artificial intelligence could generate a labeled diagram instantly. It would be accurate. It would be complete. It would not reflect the moment a student realized the veins were asymmetrical. It would not record the act of returning to look again.

When the time ends, the teacher does not ask, "Who finished?" She asks, "What did you notice after staying with it longer?" Hands rise slowly. The sketches are still imperfect. But they are attentive.

Attention here is not about silence. It is about staying long enough for detail to emerge. And detail, once noticed, changes how students see.

Illustration in the Classroom: Middle School Level

Students are analyzing two short diary entries written during the same historical event. In previous discussions, conversation often accelerated quickly. Students jumped to conclusions. Interpretations hardened within minutes. This time, the teacher adjusts the structure.

Before discussion begins, students are given five uninterrupted minutes to read and annotate silently. No devices. No conversation. Just text and pencil. At first, a few students look up early. The teacher waits. "Underline one phrase that feels important," he says. "Circle one that confuses you." The room settles into the task.

After the silent reading, discussion begins – but with a condition. Before offering an interpretation, each student must summarize the previous speaker's point. The tempo shifts. Students listen differently. They pause before responding. One student restates a peer's observation about tone and then adds, "If that's true, maybe the writer felt more afraid than angry."

The teacher notices the conversation slowing in a productive way. He resists the impulse to move them toward a "correct" interpretation. When the discussion begins to fragment, he returns them to the text. "Where do you see that?" he asks. They look down again. Nothing dramatic occurs. The class does not reach unanimous agreement. But the conversation holds complexity longer than it has before.

Artificial intelligence could summarize the diary entries in seconds. It could identify themes efficiently. It does not linger in contradiction. It does not require students to listen before speaking.

By the end of the lesson, the difference is subtle but observable. Students reference specific lines more frequently. They hesitate before dismissing an alternative reading. Attention has not been demanded. It has been structured. And in that structure, interpretation becomes less reactive and more deliberate.

Illustration in the Classroom: High School Level

Students analyze a complex case study on housing policy. The document includes data tables, stakeholder perspectives, and competing policy proposals. In previous years, discussion often accelerated quickly. Students skimmed the data, selected a position, and began debating before examining the assumptions embedded in the numbers. This time, the teacher redesigns the opening sequence. "You have eight uninterrupted minutes," she says. "No conclusions yet. Just annotate the data." The instruction is specific. Circle one data point that complicates your initial reaction. Underline a claim that feels incomplete. Write one question in the margin. The room quiets – not ceremonially, but deliberately.

A few students finish scanning quickly and glance up. The teacher remains still. She rereads a section herself, marking a column in the data table. The pause holds. When discussion begins, it begins differently. Instead of asking, "What do you think?" she asks, "What did you notice?" Students reference specific figures. One points out a discrepancy between projected cost savings and long-term maintenance expenses. Another notes that a stakeholder's testimony contradicts the data summary. The tempo slows.

At one point, a student attempts to pivot toward a firm position. The teacher gently interrupts. "Stay with the numbers a little longer," she says. The class returns to the text. No dramatic breakthrough follows. There is no consensus reached by the end of the period. But the conversation resists simplification. The teacher's design protects the tension between competing interpretations.

By the end, students are not merely defending positions. They are referencing assumptions, questioning projections, and acknowledging uncertainty. Attention here is not about silence. It is about refusing premature certainty. And that refusal – practiced consistently – strengthens judgment.

Attention is designed, protected, and explained. It is not assumed. In classrooms where attention is cultivated intentionally, students discover that depth requires time – and that time, when honored, transforms learning.

Lead with Relationship

Before any strategy works – before AI policies are clarified, before revision cycles are normalized, before attention is protected – relationship must come first. Learning is relational long before it is instructional. Students engage most deeply where they feel known.

Check in intentionally. Not as a routine attendance question, but as a genuine inquiry: "How are you arriving today?" A quick conversation at the door, a quick follow-up about yesterday's soccer game, a remembered detail about a sibling or interest – these small moments communicate something powerful: "You are not anonymous here."

Learn names quickly and use them often. Pronounce them correctly. Practice pronunciation until you get it right; do not ask for an easier abbreviation because a name is hard or new to pronounce. Ask about their origins if appropriate. Names are more than labels; they are markers of identity. In an AI age where students may compare themselves to flawless outputs and impersonal systems, hearing their name spoken with care reinforces belonging and legitimacy.

Notice effort, not just outcomes. When a student struggles publicly, the teacher's response shapes future risk-taking. A raised eyebrow can shut down participation; a curious question can open it. Instead of "That's not right," try "Tell me how you were thinking about this." Instead of "You didn't follow directions," try "Walk me through what you understood the task to be."

Curiosity preserves dignity. Judgment erodes it.

Respond to mistakes as information. When a student misuses AI, turns in shallow work, or rushes through a task, the instinct to correct quickly is understandable. But relationship reframes the moment: What need was this meeting? What pressure was this solving? What skill is still developing? When students feel safe enough to explain their choices, learning deepens.

Trust is built through consistency. Show up predictably. Follow through on commitments. Apply expectations equitably. Apologize when necessary. Students watch not only what we teach, but how we act when frustrated or uncertain. Influence grows in these quiet, repeated interactions.

In classrooms grounded in relationship, students take risks. They revise instead of hiding mistakes. They ask for help rather than outsourcing thinking. They engage with AI thoughtfully because they trust the adult guiding them. Without trust, policies feel punitive. With trust, expectations feel purposeful.

Relationship is not an "add-on" to instruction. It is the architecture that holds it together. When students believe they are seen, valued, and respected, every other strategy – feedback, questioning, revision, creativity, attention – has soil in which to grow.

Illustration in the Classroom: Elementary Level

Elementary

Each morning, the teacher stands at the doorway. She says each student's name. Not quickly. Not as roll call. Intentionally. She adjusts pronunciation when needed. She repeats a name quietly to herself if she missteps. She does not shorten names for convenience. The greetings are brief. "You looked focused during reading yesterday." "How did your sister's recital go?" "I'm glad you're here." The comments are specific. They are not performative.

One student often hesitates before entering. He avoids eye contact during math discussions and rarely volunteers answers. The teacher has noticed this pattern over several weeks. During a lesson on fractions, he attempts a new strategy and arrives at an incorrect solution. A few students glance toward him. The teacher does not correct immediately. "Tell me how you were thinking about it," she says. He explains his reasoning. Part of it is sound. One step collapses under closer examination. She names the strength first. "Your model makes sense up to here." She taps the page lightly. "What happened in this step?" He studies it again. He revises.

The correction emerges from within the conversation, not imposed from above. The class observes something important: error does not trigger dismissal. It triggers inquiry. The teacher later reflects privately that it would have been faster to state the correction directly. But speed is not the point. Preservation of dignity is.

Over time, the student raises his hand more often. Not dramatically. Not daily. But he participates without bracing. There is no visible milestone marking the change. The relationship has simply been consistent.

Relationship does not replace rigor. It makes rigor sustainable.

And because students feel known before they are evaluated, they are more willing to think aloud – even when unsure.

Middle School

Illustration in the Classroom: Middle School Level

The teacher notices a pattern. During whole-class discussion, one student rarely speaks. In small groups, she contributes thoughtfully. On written work, her reasoning is precise. Publicly, she withdraws. He does not call attention to it immediately. Instead, he begins with small consistency. At the start of class, he greets her by name. Not as spectacle. Simply as acknowledgment. "How did the robotics tournament go?" "I'm still thinking about your point yesterday about energy transfer." The references are specific. They signal memory.

During lab work, he circulates and listens before correcting. When she explains her thinking quietly to a partner, he asks, "Would you be willing to share that with the class later?" The invitation is framed as extension, not exposure. Sometimes she declines. He does not press. When she does speak in discussion, he protects the moment carefully. "What in the data led you there?" he asks – not to challenge, but to deepen. When her reasoning falters, he responds with curiosity rather than correction. "Say more about that step."

He notices that in past years, he often rushed to refine students' answers publicly. It made the discussion sharper, but sometimes narrower. He has been reconsidering that instinct. Relationship, he has come to understand, is built less through affirmation than through steadiness. After an assessment, he returns her work with two comments: one on reasoning that strengthened, one on a place where thinking could expand. The feedback is direct. It is also proportionate. There is no dramatic shift in participation.

But over time, she contributes without visible bracing. She revises without defensiveness. She asks clarifying questions during labs instead of retreating into silence.

In this classroom, relationship does not soften expectations. It stabilizes them. Because the student experiences correction as inquiry rather than exposure, intellectual risk becomes manageable. And risk – sustained, supported, consistent – becomes part of her academic posture.

Illustration in the Classroom: High School Level

A student submits essays that are analytically sharp but rarely speaks during discussion. His written margins are dense with commentary. His voice is mostly absent. The teacher notices the discrepancy. She does not call on him abruptly to correct it. Instead, she begins by naming specifics in private. "Your observation about narrative distance was precise," she says after class. "Would you be willing to bring something like that into discussion tomorrow?" The request is invitational, not corrective.

The next day, when he offers a tentative comment, she resists the impulse to refine it immediately. In earlier years, she might have tightened the phrasing or pushed quickly toward a stronger claim. Now she pauses. "What in the text led you there?" she asks. He cites a line. She nods and turns to the class. "Let's stay with that sentence." The attention shifts from him to the evidence. When his reasoning contains a gap, she does not supply the missing link. She asks another question. He revises in real time.

After discussion, she reflects privately that allowing the silence to stretch required restraint. It would have been more efficient to clarify the argument herself. But efficiency was not the goal. Over several weeks, she continues the pattern. Specific acknowledgment. Proportionate challenge. No public exposure. There is no visible turning point. He does not suddenly dominate conversation.

But he begins contributing earlier in discussions rather than waiting until the end. His comments become less guarded. He references peers' ideas and builds on them.

Relationship does not lower standards. It regulates how standards are delivered. And because critique is predictable and proportionate, participation becomes less about self-protection and more about inquiry.

This work is demanding. It is also within your reach. And it is work you are prepared to lead.

Relationship is not sentimental – it is strategic. Relationship takes patience, restraint, and courage. Those are not accidental traits. They are practices you can cultivate.

Trust creates the psychological safety necessary for risk, revision, and reflection. "In environments with high psychological safety, individuals feel comfortable taking risks, admitting errors, and offering new ideas without fear of ridicule or punishment" (Edmondson, 2019, p. 40). Trust is "the connective tissue that binds individuals together to advance the education and welfare of students" (Bryk & Schneider, 2002, p. 4). Without trust, even the most carefully designed lesson struggles to take root. With it, students engage fully and take ownership of their learning.

The work is steady, and so are you.

None of these practices require new software licenses, elaborate monitoring systems, or sweeping policy reforms. They do not depend on mastering the latest platform or anticipating the next technological trend. They require something both simpler and more demanding: clarity about purpose, consistency in practice, and courage to teach from conviction rather than fear.

Clarity means knowing what learning is for. It means understanding that the goal is not polished output, but thoughtful engagement; not speed, but depth; not compliance, but growth. When teachers are grounded in that clarity, tools – whether textbooks, tablets, or AI – fall into their proper place. They become supports, not substitutes.

Consistency means aligning daily choices with that purpose. It shows up in how questions are framed, how feedback is offered, how revision is normalized, how attention is protected, and how relationships are cultivated. Students notice patterns. They internalize what teachers repeatedly value. "Adolescents are highly sensitive to signals about what is valued in their environments" (Yeager et al., 2022, p. 585). When effort, explanation, and reflection are consistently honored, students begin to prioritize them as well.

Courage may be the most essential. Courage to slow down when the world accelerates. Courage to prioritize thinking when systems reward output. Courage to build trust instead of control. Courage to say, "Let's examine this together," rather than pretending to have every answer. Teaching in an AI age asks educators to stand firmly in what is human, even as technology evolves around them.

Teaching in this era is not about keeping up with machines. It is about staying anchored in what machines cannot replace: purpose, relationship, discernment, and meaning. When those foundations are secure, innovation becomes manageable rather than destabilizing.

On Monday morning, the work does not begin with a device or a policy. It begins with a greeting at the door. With a question posed thoughtfully. With a moment of silence that invites reflection. With feedback offered carefully. It begins, as it always has, in the space between teacher and student – where trust is built, thinking is shaped, and learning becomes human.

Chapter 19: Talking to Parents, Admins, and Communities

The work of purpose, transparency, and relationship does not end at the classroom door.

When teachers clarify why they design learning the way they do, that clarity naturally extends outward – to families, colleagues, and communities who are watching, wondering, and sometimes worrying.

For many educators, the most challenging conversations about artificial intelligence do not happen with students. They happen with adults.

One afternoon, a colleague stepped into my classroom after dismissal. He had observed from the doorway – the posted rules, the fixed desks, the quiet compliance. In his adjoining classroom, he upheld the same movement restrictions and safety protocols; the structure was not personal, it was institutional.

"How are they supposed to collaborate if they can't even move their chairs?" he said gently

He wasn't challenging me. He was pushing against the walls themselves. Beneath the question was something tender – a search for possibility inside constraint.

The facility required students to remain seated. No group work. No clustering. No movement without explicit permission. The protocols were grounded in physical safety, not pedagogy. They were written for containment, not creativity.

And I understood why.

I had experienced physical fights in that room – desks overturned, adrenaline flooding the air, bodies moving faster than thought. I knew how quickly proximity could escalate. The rules were not abstract; they were born from real harm.

Still, the tension lived in me.

I wanted students debating ideas face to face. I wanted organic collaboration. I wanted the hum of productive movement. Instead, I navigated a classroom where several adults monitored safety, where compliance was visible, and where my professional judgment sometimes felt quietly evaluated. I wore a radio clipped at my hip and a panic button on a lanyard around my neck – constant reminders that the room was governed first by physical risk, then by pedagogy.

So I told him what he could not see at first glance.

"They can't move their bodies freely," I said. "But their thinking moves constantly."

I described structured written dialogues passed from desk to desk. Socratic questioning that built on prior responses. Silent debates where students responded to posted claims in sequence. Intellectual collaboration without physical clustering.

The boundaries were real. The learning was, too.

The room was shaped by safety. It was not defined by it.

That conversation clarified something important: when adults see restriction, they often assume limitation. Our work is to make the invisible architecture of thinking visible – especially when the environment looks controlled.

Trust grows when purpose is named aloud.

That conversation was not isolated. It echoed a larger unease – one that surfaces in conference rooms, living rooms, and board meetings alike.

Parents worry about cheating, screen time, and fairness. Administrators worry about accountability, compliance, and public perception. Community members worry about whether schools are keeping up or losing control. These concerns are understandable. They come from care before cynicism. Beneath the surface of each question is a deeper desire: protect children, preserve integrity, and ensure that learning still means something.

The mistake is responding to fear with technical explanations or rigid policies alone. When a parent asks, "Will my child just copy work?" or an administrator asks, "How do we ensure equity?" the question is rarely about the mechanics of AI. They are asking something more fundamental: Can I trust this learning environment? Are students being guided wisely? Are adults still in charge of what matters?

Trust begins with shared purpose.

When educators ground conversations in values rather than tools, they shift the tone immediately. Instead of defending a platform or outlining restrictions, they articulate intentions: to teach thinking, integrity, and responsibility. To design tasks that require explanation and reflection. To ensure that students can defend their reasoning and revise their work. To prepare young people not just for assessments, but for ethical participation in a complex, technology-rich world. This reframing signals that the focus is on growth, not on catching mistakes.

Language matters more than we realize. Talking about "banning AI" frames the issue as a battle for control. Talking about "teaching responsible and thoughtful use" frames it as education. One invites resistance and suspicion. The other invites partnership and dialogue. One centers fear. The other centers formation.

With parents, clarity builds confidence. Explain how learning is assessed. Show how feedback is prioritized alongside grades. Share how assignments require students to explain their thinking, defend choices, and revise drafts. Offer concrete examples of how AI might be used transparently – and how reasoning is still evaluated. When families see that understanding cannot be outsourced, anxiety begins to soften. Parents are less concerned about tools when they understand how learning is protected. They relax when they see that students are being taught how to think, not merely how to produce.

Conversations with administrators and colleagues benefit from the same grounding. Understand and frame AI integration not as technological enthusiasm, but as pedagogical alignment. Highlight how thoughtful use supports equity by making thinking visible, how clear expectations reduce misuse, and how transparency builds professional trust. When the conversation centers on purpose, systems can evolve without sacrificing integrity.

Community trust is not built through technical mastery alone; it is built through coherence that is visible to the public. Families notice whether a school's mission statement matches classroom practice, whether the equity commitments posted on the website are reflected in curriculum choices, and whether discipline policies are applied consistently and transparently.

Trust grows when a school board presentation about student-centered learning is mirrored in classrooms where students speak, collaborate, and explain their thinking. It strengthens when grading policies shared at back-to-school night align with what appears in the online gradebook, when communication in newsletters reflects the same tone families encounter in conferences, and when community forums invite genuine dialogue rather than scripted reassurance.

Parents watch how teachers respond to emails, how principals address concerns at public meetings, and how students describe their daily experiences at the dinner table. When what educators say at open houses, on social media, in newsletters, and in district communications aligns with what students actually experience in classrooms – and when both reflect shared community values – confidence deepens.

Coherence, made visible in everyday practices, becomes the foundation of durable trust.

In an AI age, communication is part of the teaching craft. Teaching does not only happen in lesson plans; it happens in living rooms, staff meetings, and school board sessions. When educators speak from clarity rather than defensiveness, they model the very discernment they hope to cultivate in students.

Trust is not automatic. It is earned through consistency, transparency, and shared commitment to what learning is truly for.

Illustration in the Classroom: Elementary Level

At curriculum night, the teacher does not begin with tools. She begins with purpose. On the screen behind her is a simple statement: “In this classroom, we are learning to think, revise, and explain.” She shows three artifacts from a recent writing unit.

The first is a brainstorming page – handwritten ideas with a note: AI used for idea generation. The second is a draft with visible revisions: sentences crossed out, details added, margins filled with comments. The third is the final piece, paired with a short reflection: “I kept the idea about camping, but I changed most of the details because they didn’t sound like me.”

“We allow AI to help generate starting points,” she explains. “But we assess revisions, reasoning, and reflection. Students must explain what they changed and why.” She does not defend the tool. She demonstrates the process.

A parent asks, “How do you know the writing is really theirs?” The teacher gestures to the drafts. “Because I watch it develop,” she says. “I confer with students while they write. I see their revisions and hear them explain their thinking.” She pauses. “The final product matters, but I assess the thinking behind it.”

The room grows quieter – not because the concern disappears, but because the purpose is visible. Parents are not asked to trust a platform. They are invited to see the process. The assignment requires explanation. The reflection is graded. Drafts are retained. The message shared tonight mirrors what students hear daily. Nothing dramatic occurs. But families leave with clarity – and confidence.

Illustration in the Classroom: Middle School Level

At a staff meeting, a discussion about AI turns tense.
Several teachers express concern about academic dishonesty.
A principal asks how grading can remain reliable if students have access to generative tools at home. Instead of responding with reassurance alone, one teacher shares a recent unit plan. She projects the assignment sequence.

Students were required to: Submit initial annotations of a historical article. Identify where an AI-generated summary oversimplified the author's claim. Write a short analysis comparing the summary to the original text. Attach a reflection explaining which interpretations they accepted, revised, or rejected. She does not describe the unit as innovative. She describes it as aligned. "Our focus remains on interpretation and reasoning," she says. "AI is used openly – but it cannot complete the comparison, the critique, or the reflection."

She shows student samples. In one, a student highlights where the summary flattened nuance. In another, a student explains why a phrase in the original text carried more urgency than the AI version. The principal studies the reflection rubric. "You're grading the reasoning, not just the product," she says. "Yes. The final paragraph matters. But the analysis of the summary matters more." the teacher replies. The room shifts – not emotionally, but analytically. The conversation moves from "How do we prevent misuse?" to "How do we structure assignments so reasoning is required?"

No policy is rewritten that afternoon. But something clarifies. AI is not being integrated casually, but rather it is being contained within instructional purpose. Trust grows not because the tool is defended – but because the design is coherent. And coherence, once visible, reduces suspicion.

Illustration in the Classroom: High School Level

A school board meeting includes discussion about academic integrity. A board member asks directly, "How do we know students aren't just letting AI do the work?" The teacher with structure, not policy. She distributes a recent assignment from her civics class. The sequence is clear: Students generated an AI summary of a public policy position. They identified omissions and oversimplifications. They compared the summary to primary sources. They submitted a written reflection documenting where their thinking shifted. She projects excerpts from student reflections.

One student writes: "The AI summary made the policy sound neutral, but the speech we read used emotionally charged language. That changed how I interpreted the intent." Another notes: "I agreed with the summary, but after reviewing the data table, I revised my position." The teacher explains how the rubric weights reasoning, source evaluation, and revision more heavily than fluency. "We do not assess whether students can produce polished language," she says. "We assess whether they can defend their interpretation."

A board member asks whether misuse still occurs. "Yes," she replies. "Which is why assignments require explanation. If students cannot articulate how they arrived at a conclusion, the work is incomplete."

There is no dramatic moment. No applause. No sweeping shifts. But the questions become more precise. "How do we support this kind of assessment across departments?" "What professional development would help teachers design similar sequences?" The conversation moves from surveillance to structure. Trust grows because the learning process is visible and defensible. Coherence – between state purpose, assignment design, and assessment practice – becomes the foundation for confidence.

With administrators, alignment matters. Frame conversations about AI within shared priorities: instructional quality, equity, student growth, and professional judgment. Emphasize what you believe about thoughtful integration reducing misuse, increasing engagement, and strengthening – not weakening – teacher expertise. Leaders are far more likely to support approaches that are coherent and purpose-driven rather than reactive or fear-based. Be honest about your intent, your current capacity, and your focus. Credibility grows when educators acknowledge both possibility and limits.

With communities, storytelling is powerful. Policies describe boundaries; stories reveal impact. Share classroom moments that illustrate curiosity, struggle, revision, and growth. These narratives humanize learning in ways technical explanations never can. They show that thinking – not shortcuts – remains at the center.

Humility builds credibility across all of these conversations. No educator has everything figured out, and pretending otherwise erodes trust. Acknowledging uncertainty while demonstrating clear purpose fosters confidence. Saying, "We are learning alongside our students," signals leadership rather than weakness. It communicates commitment to guidance, reflection, and growth – not surveillance or arbitrary control.

When adults approach AI with anxiety, that fear does not remain private; it settles into policy and seeps into classroom culture. Reactive restrictions may feel decisive, yet they often deepen the very mistrust they seek to contain. Research across leadership and educational technology warns that "low psychological safety ... increases defensive decisions, whereas psychological safety mitigates defensive, fear-based responses" (Artinger et al., 2025), and that "a lack of trust increases anxiety ... which inhibits technology adoption" (Shen et al., 2025).

Fear narrows vision. Dialogue widens it. Thoughtful conversation clarifies intention, restores perspective, and replaces surveillance with shared responsibility. In climates shaped by trust rather than threat, innovation becomes reflective instead of reactive, and policy becomes a tool for learning rather than a response to alarm.

Boundaries still matter. Responsible AI use does not mean unlimited or unexamined use.

Expectations, transparency, and accountability remain essential – but the rationale shifts. Boundaries exist to protect learning, preserve integrity, and support development – not simply to punish misuse. When parents, administrators, and community members understand that purpose, fear diminishes and partnership strengthens.

Education has always been relational. AI does not change that; it makes it more visible.

The work of teaching extends beyond content delivery into relationships with families, colleagues, and the broader community. When educators communicate with clarity, transparency, and care, they cultivate trust – the foundation on which meaningful learning rests.

When communities experience "transparent communication and authentic engagement," trust strengthens and fear recedes (Darling-Hammond et al., 2020, p. 41).

And when educators communicate with clarity and care, they build the collective confidence necessary for students – and the adults who support them – to thrive in an AI-rich world.

Chapter 20: Taking Care of the Teacher

As conversations expand beyond the classroom – to families, administrators, and communities – the scope of teaching becomes more visible. Educators are not only designing lessons and guiding students; they are explaining, reassuring, adapting, and leading. That outward-facing work matters. It builds trust and strengthens partnership. But it also carries weight.

Sustaining others requires tending to the one doing the sustaining. The well must be replenished if it is to keep giving water. The flame that lights the room must be shielded from the wind. Burnout is not devotion, it is depletion. Sustainability is an ethical responsibility, not indulgence.

Change is exhausting ... even when it is meaningful.

Teachers are asked to adapt constantly: new standards, new tools, new expectations, new crises. Artificial intelligence arrives not as a single initiative, but as a shifting presence – altering how work is completed, how learning is measured, how integrity is defined, and even how expertise is perceived. It reshapes lesson planning, assessment, feedback, communication, and professional identity. The ground feels as if it is always moving.

It is reasonable to feel tired.

At the end of one long day, I sat alone in my classroom after dismissal. The room had quieted. The fluorescent lights hummed softly overhead. I reached down and unclipped the radio from my waistband, then lifted the panic button from around my neck. The weight of them had become so familiar I hardly noticed it during the day – until I took them off.

For a moment, I just held them in my hands.

Those devices were there for safety. They were necessary. I had used the radio before. I had felt the surge of adrenaline when a situation escalated faster than words could slow it. Hypervigilance was not theoretical in that setting; it was embodied.

And yet, after removing them, I looked at my desktop.

An email waited about AI integration. Opportunity to participate in district adoption loomed. A planning document needed revision. A new tool required exploration.

The irony was not lost on me. My students did not have open internet access. For safety reasons, their digital world was tightly restricted, filtered, and supervised. They could not experiment freely with emerging platforms. They could not wander through the vastness of information the way other students might. And yet I felt the steady pressure to remain current – to understand what was shaping the broader educational landscape, to prepare them for a world they could not fully access within those walls.

And I knew that each of them would one day step back into that world; in Washington, juvenile confinement did not extend beyond young adulthood.

I was holding two realities at once.

Inside the classroom, I navigated radios, panic buttons, and movement protocols designed for containment. Outside it, I navigated webinars, policy updates, and conversations about artificial intelligence transforming instruction. My students lived within constraint. The profession demanded fluency.

The distance between those two worlds felt wider in the quiet of that room.

Different demands. Same depletion.

And I remember wondering how to honor both – how to remain informed without becoming consumed, how to bring thoughtful, relevant learning to students whose access was limited, and how to do so without fraying the very steadiness they relied on.

The room was silent. My shoulders were not.

That evening, I closed the laptop earlier than usual. Not out of neglect, but out of recognition.

Sustainability is not withdrawal from the work. It is the decision to remain whole enough to continue it.

Taking care of the teacher is not a personal indulgence. It is a professional necessity. Teachers flourish not in isolation or pressure, but where there is "coherent support, collective efficacy, and a culture that honors slow thinking, reflective practice, and mutual care" (Fullan & Langworthy, 2014, p. 26). When systems protect time for reflection, normalize collaborative problem-solving, and signal that adults matter as much as students, the work remains sustainable.

When expectations expand but time and recognition do not, exhaustion becomes predictable.

AI will not solve this problem. If anything, it risks accelerating it by increasing pace and pressure. Efficiency can create the illusion that more can be done in less time – more feedback, more communication, more documentation, more responsiveness. But faster does not always mean healthier. More tools do not automatically mean less strain.

What sustains teachers is not technology, but meaning.

Teachers endure when they can see the impact of their work – when relationships deepen, when students grow, when insight sparks across a classroom. They thrive in environments that value reflection over reaction and growth over perfection. They remain resilient when professional judgment is trusted rather than constantly questioned.

This means setting boundaries.

Not every tool needs to be mastered. Not every platform needs to be adopted. Not every new initiative requires instant compliance. It is okay to pause, observe, experiment thoughtfully, and decide intentionally what aligns with your values and your students' needs. You may need to advocate for yourself and your students, and that advocacy matters. You are your student's best advocate.

Discernment applies to adults as much as to students. It also means reclaiming identity.

Teachers are not customer service representatives for systems that never sleep. They are skilled professionals engaged in relational, cognitive, and ethical work. They are designers of learning, cultivators of attention, guardians of belonging, stewards of creativity, and guides of moral reasoning. Naming that work matters. Remembering its depth matters.

Community is part of care.

Teaching has always been sustained by shared experience: hallway conversations that normalize struggle, collaborative planning that lightens load, shared laughter that releases tension, collective problem-solving that restores hope. Isolation increases fatigue. Connection restores perspective; strong collegial relationships "were significantly associated with greater teacher well-being" (Blair et al., 2024, p. 501). In an AI age that can feel impersonal and accelerated, human collegiality becomes even more essential.

Taking care of the teacher is not separate from taking care of students.

When educators are supported and grounded in purpose, they teach with clarity rather than urgency. They respond with patience rather than depletion. They model the very balance, discernment, and reflection they hope students will develop.

Sustaining the profession requires sustaining the people within it.

And that work begins with permission: to slow down, to set limits, to seek support, and to remember why you chose this work in the first place.

Illustration in the Classroom: Elementary Level

A teacher begins noticing subtle shifts in her classroom.

Students seem more distracted and increasingly frustrated. Between new content, digital platforms, and AI-supported practice apps, the pace has accelerated. She finds herself moving quickly from table to table, answering questions efficiently but feeling mentally scattered. The tools are functioning. The learning feels thin.

Instead of interpreting her fatigue as personal inadequacy, she treats it as information. She simplifies the structure of math block – reducing the number of rotations and limiting the digital component to one purposeful task. She adds a brief five-minute reflection circle at the end of class where students name what strategy helped them and what still feels confusing. She collaborates with a grade-level colleague to share planning for AI-supported activities, distributing the cognitive load. She communicates the shift to families, explaining that depth and clarity matter more than coverage.

As the pace steadies, the atmosphere changes. Students linger longer with multi-step problems. They explain their reasoning more thoughtfully and show less visible frustration. The teacher feels more present, less reactive. When she encounters a challenging question, she models calm problem-solving aloud, demonstrating how to pause, reconsider, and proceed deliberately.

The adjustment is not dramatic, but it is intentional. By protecting her own capacity and designing with sustainability in mind, she creates conditions where students can think more deeply. Her well-being becomes part of the instructional design.

In this classroom, care for the teacher is not separate from care for students. It is foundational to it.

Illustration in the Classroom: Middle School Level

Students are energized by the immediacy of an AI-assisted lab.

It simulates chemical reactions and generates data visualizations in seconds. Despite student excitement, their teacher feels a quieter pressure building – to master every feature, troubleshoot technical glitches, keep pace with curriculum maps, and safeguard academic integrity. He notices himself staying later each afternoon, revising lessons to "keep up." The innovation is exciting, but the strain is real.

Rather than sustaining that pace indefinitely, he establishes boundaries. Certain lab days are designated for AI-supported simulations; others are intentionally unplugged, centered on hands-on experimentation and discussion. He tells students plainly, "We don't need every feature to learn deeply." He limits his after-hours planning time and creates a shared resource folder so colleagues can contribute materials, distributing the workload rather than absorbing it alone.

He also builds in regular department conversations focused not only on student data, but on teacher capacity. They discuss what worked, what felt overwhelming, and what can be simplified. These exchanges normalize uncertainty. The pressure to appear fluent in every emerging tool softens when collaboration replaces isolation.

With steadier energy, his instruction shifts. He notices subtle misconceptions that simulations gloss over. He asks more probing questions. He listens more attentively during lab discussions. Students sense the difference – a teacher who is thoughtful and present rather than hurried and reactive.

By honoring his limits, he protects his effectiveness. The integration of AI becomes intentional rather than compulsive. In this classroom, care for the teacher is not a retreat from rigor; it is what makes sustained rigor possible.

Illustration in the Classroom: High School Level

AI tools are integrated. They help students trace thematic connections across novels and compare interpretations efficiently. Discussions grow more layered as students arrive with annotated drafts and synthesized research. Yet behind the scenes, the workload expands: reviewing AI-assisted essays, facilitating Socratic dialogue, conferencing individually, and navigating nuanced questions about authorship and intellectual integrity. The teacher begins to feel the strain – mentally taxed, emotionally stretched, and perpetually "on." The classroom hums with depth, but sustaining that depth feels increasingly heavy.

Instead of absorbing the pressure silently, she recalibrates. She reduces the number of formal essays and increases the quality of feedback on fewer, more substantial pieces. She blocks protected time in her schedule for uninterrupted reading and response, resisting the urge to multitask. She tells students candidly, "If I'm going to challenge your thinking well, I need space to think carefully about it." The transparency reframes pacing as intentional rather than indulgent.

She also models discernment with technology. When asked whether she plans to experiment with every new AI writing feature, she answers honestly: "I explore what strengthens our learning. I don't need every tool to do good work." In doing so, she demonstrates that innovation is a choice, not an obligation.

During individual conferences, her presence is steady. She listens closely, asks clarifying questions, and names patterns in reasoning that students might not see themselves. The conversations feel deliberate rather than rushed. Students leave not only with clearer revisions, but with a sense that their ideas were taken seriously. By protecting her own capacity, she preserves the relational core of her classroom.

The rigor remains – but it is sustainable. In an environment shaped by rapid technological change, she models a different kind of strength: steadiness.

Illustrations from the Classroom: University Level

In an AI-saturated academic climate, I found myself exhausted in a way I had not anticipated. Every paper felt suspect. Was this paragraph written by the student? Was this synthesis earned or generated? Was I grading understanding – or editing machine output? I noticed something subtle happening in me. I was reading defensively. Instead of engaging ideas, I was scanning for irregularities. Instead of responding to reasoning, I was looking for seams. The posture was unsustainable.

One evening, after hours of reviewing polished but hollow submissions, I realized the problem was not simply academic integrity. It was design. I had structured assignments in ways that rewarded polish. So, I recalibrated.

I reduced the number of written submissions and increased live dialogue. Major papers were accompanied by recorded reflections explaining how ideas evolved. Oral defenses became routine, not punitive. I built protected time into my schedule for uninterrupted reading – not grading between meetings, not responding at midnight. And I told my students the truth. "Thoughtful feedback takes time," I said. "I want to give you my best thinking. That means we will do fewer things more deeply." The relief in the room surprised me.

They, too, were exhausted – navigating when to use AI, how to use it, whether using it diminished their growth. The surveillance posture I had felt was mirrored by their anxiety. So, we named it. We discussed tool boundaries openly. We clarified when AI could assist brainstorming and when it would short-circuit learning. We distinguished between support and substitution. Something shifted. Grading felt less adversarial. Discussion regained texture. My energy stabilized. The work felt human again.

Artificial intelligence accelerates productivity. It also tempts overextension. It invites teachers to monitor more, produce more, respond faster. But sustainability does not come from acceleration; it comes from boundaries.

Taking care of the teacher in an AI world is not retreat. It is discernment. It is knowing that vigilance cannot be your primary stance. It is designing for integrity rather than policing for deception. It is protecting your cognitive bandwidth so you can remain present.

At the university level – as in every other classroom – the art of teaching depends on a teacher who is not depleted beyond reflection.

Care for students and care for self are not competing priorities. They are interdependent. Technology may change our tools. It does not change our limits.

And in a profession built on intellectual presence, preserving that presence is not indulgence. It is responsibility.

Self-care is often framed as an individual responsibility – take a break, set boundaries, practice mindfulness. While these practices matter, they are not sufficient on their own. Care must also be structural. "Teacher stress is shaped not only by individual coping skills but by the organizational conditions in which teachers work" (Greenberg et al., 2016, p. 2).

Teachers need time built into schedules for planning and reflection. They need trust in their professional judgment rather than micromanaging surveillance. They need autonomy to decide which tools serve learning and which can wait. And they need leaders who protect space for deep work instead of layering initiative upon initiative without pause. They also need permission to be learners themselves.

In an AI-rich landscape, no educator will have every answer. Schools that normalize experimentation, shared problem-solving, and gradual integration create room for growth without shame. When teachers are allowed to say, "I'm still figuring this out," they remain curious instead of defensive. That posture sustains innovation far more effectively than pressure ever could. Above all, teachers need reassurance that their work remains essential – and it truly does.

AI may change how education looks, but it does not replace the human presence at its center. Students do not learn because systems are efficient or platforms are optimized. They learn because someone notices their confusion, celebrates their progress, and refuses to give up on them. They learn because a teacher asks a harder question, offers a steadier hand, or believes in their capacity before they believe in themselves.

Teaching is not a race against technology. It is a commitment to humanity. Caring for oneself in this profession is not stepping back from responsibility; it is preserving the energy required to carry it well. It is choosing sustainability over burnout, depth over frenzy, and purpose over pressure. When teachers are supported, trusted, and sustained, classrooms become places of steadiness in a world that often feels accelerated and unstable.

The future of learning depends on teachers who are not only skilled, but whole. In the age of intelligent machines, the art of teaching endures – not in delivering information faster than a device, but in cultivating understanding, guiding judgment, shaping identity, and nurturing the human spirit.

EPILOGUE: THE FUTURE NEEDS TEACHERS

The classroom looks different now.

Screens are more common. Answers arrive instantly. Students navigate a world where information is abundant and certainty is often simulated. Artificial intelligence is woven into daily life in ways that feel both ordinary and astonishing. What once required hours of searching can be generated in seconds. What once demanded memorization can be retrieved with a prompt. And yet, the heart of learning remains unchanged.

Across this shifting landscape, we return to what endures. Answers matter less than the questions that shape them, reflection deepens what performance alone cannot sustain, and influence carries farther than authority. Feedback forms identity more powerfully than grades, discernment outlasts speed, and innovation must be anchored in ethics if it is to serve rather than distort learning.

Protection of attention in an age designed to fracture it, cultivation of belonging so that voice becomes possible, and honoring of creativity as something no algorithm can replicate are foundational to the art of teaching. Trust stretches beyond classroom walls into families and communities, and sustaining teachers is inseparable from sustaining learning itself.

In truth, tools evolve, but the human foundations of teaching endure. As the tools, the human work of teaching does not disappear. It becomes clearer.

Students still need someone to notice when confusion hides behind confidence. They still need spaces where questions are welcomed and mistakes are safe. They still need adults who can help them make sense of complexity, contradiction, and possibility. They need someone to ask, "Why do you think that?" and then wait for the answer. They need someone to say, "Try again," and mean it as an invitation, not a verdict.

Technology has never replaced that work. It has only made it more visible. When answers are abundant, discernment is rare. When output is instant, ownership matters. When language is machine-polished, authentic voice becomes courage.

The future does not need teachers who can compete with machines. It needs teachers who can do what machines cannot: cultivate judgment, nurture belonging, protect attention, invite revision, model humility, and guide ethical reasoning. It needs educators who understand that learning is not about speed, but about meaning. Not about performance, but about formation.

Teaching has always been an act of hope. Every lesson assumes a future worth preparing for. Every question invites a learner to imagine themselves as capable of growth. Every moment of feedback says, "You are not finished yet." Every act of patience declares that development takes time.

Artificial intelligence may reshape the tools of education, but it does not redefine its purpose. The purpose remains deeply human: to help learners become thoughtful, creative, ethical participants in a shared world. To help them ask better questions. To help them evaluate what they encounter. To help them create rather than copy. To help them belong without erasing who they are. To help them use powerful tools wisely, rather than be used by them.

The future will belong to those who can think critically, attend deeply, create courageously, and act responsibly. These capacities are not programmed. They are cultivated – patiently, relationally, and intentionally. They grow in classrooms where teachers protect reflection, celebrate revision, and treat learning as an unfolding process rather than a finished product.

Teachers are not relics of a pre-digital past. They are architects of a humane future. They design experiences that make thinking visible. They create communities where voice is legitimate. They model integrity in the face of ambiguity. They remind students that intelligence is not measured by speed, but by depth.

As classrooms continue to evolve, one truth remains: learning happens where humans connect. Where curiosity is protected. Where creativity is invited. Where mistakes are reframed as information. Where trust allows risk. Where someone says, "Your thinking matters because it is yours."

The future needs teachers – not because machines have failed, but because humanity still matters. And it always will.

REFERENCES

Artinger, F. M., Marx-Fleck, S., Junker, N. M., Gigerenzer, G., & van Dick, R. (2025). *Coping with uncertainty: The interaction of psychological safety and authentic leadership in their effects on defensive decision making. Journal of Business Research.* Advance online publication. https://doi.org/10.1016/j.jbusres.2025.115240

Bandura, A. (1997). *Self-efficacy: The exercise of control.* W. H. Freeman.

Bandura, A. (2001). Social cognitive theory: An agentic perspective. *Annual Review of Psychology, 52*(1), 1–26. https://doi.org/10.1146/annurev.psych.52.1.1

Blair, E. E., Smith, J., & Jones, A. (2024). *Teacher collegial relationships and well-being during the COVID-19 pandemic. School Psychology Quarterly.* https://doi.org/10.1037/spq0000596

Blum, R. W. (2005). *Health and academic achievement: Strengthening the connection.* University of Minnesota, Center for Adolescent Health and Development.

Bransford, J. D., Brown, A. L., & Cocking, R. R. (2000). *How people learn: Brain, mind, experience, and school.* National Academy Press.

Broussard, M. (2018). *Artificial unintelligence: How computers misunderstand the world.* MIT Press.

Bryk, A. S., & Schneider, B. (2002). *Trust in schools: A core resource for improvement.* Russell Sage Foundation.

Cremin, L. A. (1957). *The American common school: An historic conception*. Teachers College Press.

Darling-Hammond, L., Flook, L., Cook-Harvey, C., Barron, B., & Osher, D. (2020). Implications for educational practice of the science of learning and development. *Applied Developmental Science, 24*(2), 97–140. https://doi.org/10.1080/10888691.2018.1537791

Davis, E., & Marcus, G. (2023). Commonsense reasoning and large language models. *Communications of the ACM*, 66(10), 6–8.

Deci, E. L., & Ryan, R. M. (2000). The "what" and "why" of goal pursuits: Human needs and the self-determination of behavior. *Psychological Inquiry, 11*(4), 227–268. https://doi.org/10.1207/S15327965PLI1104_01

Dennett, D. (1987). *The intentional stance.* MIT Press.

Dweck, C. S. (2006). *Mindset: The new psychology of success.* Random House.

Edmondson, A. C. (2019). *The fearless organization: Creating psychological safety in the workplace for learning, innovation, and growth*. Wiley.

Feiman-Nemser, S. (2001). From preparation to practice: Designing a continuum to strengthen and sustain teaching. *Teachers College Record, 103*(6), 1013–1055.

Fullan, M., & Langworthy, M. (2014). *A rich seam: How new pedagogies find deep learning*. Pearson.

Goldin, C. (1999). *A brief history of education in the United States* (NBER Historical Paper No. 119). National Bureau of Economic Research. https://doi.org/10.3386/h0119

Greenberg, M. T., Brown, J. L., & Abenavoli, R. M. (2016). Teacher stress and health: Effects on teachers, students, and schools. *Review of Educational Research, 86*(4), 1–38. https://doi.org/10.3102/0034654316629250

Hammond, Z. (2015). *Culturally responsive teaching and the brain: Promoting authentic engagement and rigor among culturally and linguistically diverse students.* Corwin.

Hanford, E. (Host). (2022). *Sold a story* [Audio podcast]. American Public Media. https://features.apmreports.org/sold-a-story/

Hatano, G., & Inagaki, K. (1986). Two courses of expertise. *Child Development, 57*(2), 262–272. https://doi.org/10.2307/1130593

Hattie, J. (2009). *Visible learning: A synthesis of over 800 meta-analyses relating to achievement.* Routledge.

Hawkins, J. (2023). *On intelligence and prediction in AI systems.* MIT Press.

Himmele, P., & Himmele, W. (2017). *Total participation techniques: Making every student an active learner* (2nd ed.). ASCD.

Immordino-Yang, M. H. (2016). *Emotions, learning, and the brain: Exploring the educational implications of affective neuroscience.* W. W. Norton & Company.

Immordino-Yang, M. H., & Damasio, A. (2007). We feel, therefore we learn: The relevance of affective and social neuroscience to education. *Mind, Brain, and Education, 1*(1), 3–10. https://doi.org/10.1111/j.1751-228X.2007.00004

Immordino-Yang, M. H., Darling-Hammond, L., & Krone, C. R. (2019). *The brain basis for integrated social, emotional, and academic development*. Aspen Institute.

Kaestle, C. F. (1983). *Pillars of the republic: Common schools and American society, 1780–1860*. Hill and Wang.

Knoster, T., Villa, R. A., & Thousand, J. S. (2000). *A framework for thinking about systems change*. In R. A. Villa & J. S. Thousand (Eds.), *Restructuring for caring and effective education: Piecing the puzzle together* (pp. 93–128). Paul H. Brookes Publishing.

Lang, P., & Metcalf, H. (2021). *The slow professor: Challenging the culture of speed in the academy* (2nd ed.). University of Toronto Press.

Liljedahl, P. (2021). *Building thinking classrooms in mathematics: 14 teaching practices for enhancing learning*. Corwin.

Mahbodi, M., Papi, M., & Wolff, D. (2025). *How learners' trust in teachers shapes their feedback-seeking behaviors. Language Teaching Research*. Advance online publication. https://doi.org/10.1177/13621688251352268

Mollick, E. (2023). *Co-intelligence: Living and working with AI*. Portfolio/Penguin.

Murphy, M. C. (2023). *Cultures of growth: How the new science of mindset can transform individuals, teams, and organizations*. Simon & Schuster.

National Research Council. (2012). *Education for life and work: Developing transferable knowledge and skills in the 21st century*. The National Academies Press. https://doi.org/10.17226/13398

Organisation for Economic Co-operation and Development. (2025). *Seven questions about creativity and creative thinking: Insights from PISA 2022*. OECD Publishing.

Pekrun, R. (2006). The control-value theory of achievement emotions: Assumptions, corollaries, and implications for educational research and practice. *Educational Psychology Review, 18*(4), 315–341. https://doi.org/10.1007/s10648-006-9029-9

Perkins, D. N. (1992). *Smart schools: From training memories to educating minds*. Free Press.

Pianta, R. C., Hamre, B. K., & Allen, J. P. (2012). Teacher-student relationships and engagement: Conceptualizing, measuring, and improving the capacity of classroom interactions. In S. L. Christenson, A. L. Reschly, & C. Wylie (Eds.), *Handbook of research on student engagement* (pp. 365–386). Springer.

Popham, W. J. (2001). *The truth about testing: An educator's call to action*. ASCD.

Reeve, J. (2012). A self-determination theory perspective on student engagement. *Handbook of Research on Student Engagement*, 149–172. Springer. https://doi.org/10.1007/978-1-4614-2018-7_7

Ritchhart, R., Church, M., & Morrison, K. (2011). *Making thinking visible: How to promote engagement, understanding, and independence for all learners*. Jossey-Bass.

Rogoff, B. (2003). *The cultural nature of human development*. Oxford University Press.

Russell, S., & Norvig, P. (2021). *Artificial intelligence: A modern approach* (4th ed.). Pearson.

Safir, S., & Watson, J. (2021). *Street data: A next-generation model for equity, pedagogy, and school transformation*. Corwin.

Schunk, D. H., Pintrich, P. R., & Meece, J. L. (2008). *Motivation in education: Theory, research, and applications* (3rd ed.). Pearson.

Schunk, D. H., & DiBenedetto, M. K. (2020). Motivation and social-emotional learning: Theory, research, and practice. *Contemporary Educational Psychology, 60*, 101821. https://doi.org/10.1016/j.cedpsych.2019.101821

Shanahan, M. (2024). Talking about large language models. *Communications of the ACM*, 67(2), 14–17.

Shen, L., *et al.* (2025). *Psychological safety and trust as drivers of teachers' continued use of AI tools in classrooms. Scientific Reports*. Advance online publication.

Smith, D., Fisher, D., & Frey, N. (2015). *Better than carrots or sticks: Restorative practices for positive classroom management*. ASCD.

Sugai, G., Horner, R., Dunlap, G., Hieneman, M., Lewis, T., Nelson, C., Scott, T., Liaupsin, C., Sailor, W., Turnbull, A., Turnbull, R., Wickham, D., Wilcox, B., & Ruef, M. (2000). Applying positive behavior support and functional behavioral assessment in schools. *Journal of Positive Behavior Interventions, 2*(3), 131–143. https://doi.org/10.1177/109830070000200302

Tyack, D., & Cuban, L. (1995). *Tinkering toward utopia: A century of public school reform*. Harvard University Press.

Van Allen, Z., & Freedberg, M. (2024). *Adaptive learning systems and organizational change. Journal of Educational Technology & Society, 27*(1), 45–60.

Vygotsky, L. S. (1978). *Mind in society: The development of higher psychological processes.* Harvard University Press.

Wentzel, K. R. (2010). Students' relationships with teachers as motivators of engagement in school: Implications for theory, research, and practice. In T. C. Urdan & S. A. Karabenick (Eds.), *Advances in motivation and achievement: Vol. 16A. The decade ahead: Theoretical perspectives on motivation and achievement* (pp. 143–187). Emerald Group Publishing.

Yeager, D. S., & Dweck, C. S. (2012). Mindsets that promote resilience: When students believe that personal characteristics can be developed. *Educational Psychologist, 47*(4), 302–314. https://doi.org/10.1080/00461520.2012.722805

Yeager, D. S. (2022). *10 to 25: The science of motivating young people*. Atria Books.

Yeager, D. S., Bryan, C. J., Gross, J. J., Murray, J. S., Krettek Cobb, D., Santos, P. H., Gravelding, H., Johnson, M., Jamieson, J. P., & Walton, G. M. (2022). A synergistic mindsets intervention protects adolescents from stress. *Nature, 607*(7919), 512–520. https://doi.org/10.1038/s41586-022-04907-7

Zins, J. E., Weissberg, R. P., Wang, M. C., & Walberg, H. J. (2004). *Building academic success on social and emotional learning: What does the research say?* Teachers College Press.

Made in the USA
Monee, IL
18 March 2026

45821758R00164